Climate and Human Variability

Second Edition

D. F. Roberts
The University of Newcastle upon Tyne

Cummings Publishing Company
Menlo Park, California • Reading, Massachusetts
London • Amsterdam • Don Mills, Ontario • Sydney

This book is in the
Cummings Modular Program in Anthropology

Cover photo: Shilluk of the Sudan grasslands—
extreme heat, linear physique.

Printed in the United States of America.
Published simultaneously in Canada.
Library of Congress Catalog Card No. 76-2227

ISBN 0-8465-6625-7
abcdefghijk-AL-798

Cummings Publishing Company, Inc.
2727 Sand Hill Road
Menlo Park, California 94025

Preface

Climate and Human Variability draws together the evidence relating physical variations in humans to the climates in which they live. Empirical data from physiological and anthropometric studies are consistent in showing that adult body form varies from one population to another in such a way as to confer some advantage in the environment inhabited. The increasing data from experimental physiology support this hypothesis. The evidence on the way in which the climatic association comes about suggests that it is genetic differences that are largely responsible, through natural selection acting on the genepools of populations. There are far-reaching implications of such climatic preadaptation for the great cultural advances of human history.

This book is founded on a wide reading of the classical literature of physical anthropology interpreted in the light of more modern adaptive physiology and evolutionary genetics. It uses and places in perspective the large number of descriptive anthropometric studies of human groups that have been made in the past, and suggests that climatic adaptation of body form may be among the best established examples of selection for quantitative characters in humans.

Climate and Human Variability is a useful supplement to courses in physical anthropology, human ecology and evolution, adaptive physiology, human geography, and human population genetics, and it will be of philosophical interest to all concerned with the evolutionary emancipation of humans from their environment.

About the Author

Derek F. Roberts is Head of the Department of Human Genetics in the University of Newcastle upon Tyne, a department which he initiated in 1966 and subsequently developed to its present status of a major center of human genetic work in Britain. He holds the degree of D.Phil. from the University of Oxford, and M.A. and Sc.D. from the University of Cambridge. In 1949 he joined the Department of Human Anatomy, Oxford, and during his years there he examined his ecological and genetic ideas in fieldwork carried out among African grassland peoples. These and other field studies have provided ample material for his numerous writings. Much of Roberts' effort since moving to Newcastle has been devoted to running his research and service department, but he still finds time for heavy research commitments and takes an active interest in the affairs of several learned societies. He is currently President of the Clinical Genetics Society of Great Britain, and Vice-Chairman of the Society for the Study of Human Biology.

Contents

Introduction

In these days of world travel and long-distance vacations, the variations in climate that are encountered from place to place are no longer surprising. Yet before the present century there were relatively few individuals who had themselves experienced climatic differences; the majority of mankind knew of them only by hearsay, often embroidered, and were personally acquainted only with changes in weather, those day-to-day and month-to-month fluctuations in temperature, humidity, insolation, precipitation, and windspeed around a mean level that in sum make up the climate of a place. It was therefore not surprising that the possibility that climate was an influence on human form, when it was considered at all, remained essentially speculative.

The idea of association between man's physical features and his habitat is not new. Hippocratic medicine endeavored always to see man against the background of his environment, and the treatise on "airs, waters, places" *(De aere, aquis, et locis)* attributed to this school flourishing in the fifth century B.C. on the island of Cos, attempted to explain observed physical differences in terms of climate, the nature of the country inhabited, and numerous other environmental factors. Aristotle, in his *Historia Animalium*, contributed observations on regular correspondence between the coloring of people and the latitude in which they lived. Poseidonius thought that a hot climate was responsible for skin color, flatness of nose, thickness of lips, and frizzled hair, but noted that Indians, although living in the same latitude as Ethiopians, exhibited considerable physical differences from them. Since those early days of science, the problem has exercised the minds of scholars too numerous to list, though a few of the most interesting individuals may be picked out. Fascinating today are the efforts of the early nineteenth century—those of Stanhope Smith (1810), for example, with his postulated effects of cold environments on physiognomy (severe cold has the effect of contracting the eyebrows and eyelids, closing the mouth, and raising the cheeks, and this has produced the short broad face of the Tartars); or of de Müller (1853) (in hot climates the amount of oxygen inspired is insufficient to change the carbon to carbonic acid and the unconsumed carbon is deposited in the pigment cells of the skin to give the Negro pigmentation). The "founder of physical anthropology," Blumenbach (1752-1840), pointed the way for development with his recognition that perpetuated variations in animals might arise in response to external influences. Yet even Darwin's work (1871, 1875), though in other respects revolutionizing the study of mankind, failed to produce any investigation of the geographical distribution of existing human characteristics—as a preliminary to the study of their evolutionary significance—perhaps because Darwin himself was less impressed with the geographic regularity of human variation than he was with the similarities between Indians of the Amazon and Tierra del Fuego despite their very different environments.

It was not until the present century that studies began to be made of the effect of environment on morphological features. Boas' pioneer study (1910) suggested changes of head form in descendants of immigrants to the United States, a proposal received with skepticism and shown to be based on statistically insufficient evidence. However, Boas performed further investigations in Central America which seemed to confirm his earlier work. Shapiro (1939) demonstrated corroboratory divergences between Japanese migrants to Hawaii and their relatives remaining in Japan. Ivanowsky (1925) investigated the effects of undernutrition and found extensive morphological changes. Marrett (1936) wrote, if somewhat uncritically, on the effects of different environments on human evolution, with emphasis on variations in chemical availability. These and a few other suggestions notwithstanding—very few, in respect of the time factor—with the notable exception of the work of Thomson and Buxton (1923) on nasal shape, objective investigation of the relationships of human form to climate did not begin to appear until after the 1939-1945 war.

Is any association to be expected between human physical characteristics and environment? Apart from these hints, other fields of study provide a certain amount of evidence. Paleontology has suggested that adaptation to environment, by the mechanism of natural selection acting on the genetics of populations, is at least the major nonrandom influence on the direction of biological evolution (the principle is elegantly stated by Simpson, 1950). Zoology has revealed the inaccuracy of the concept of the fixity of animal groups, and it has indicated that much of the differentiation observed among such groups is directly attributable to environmental adaptation. A number of so-called laws have been formulated to express some of the relationships of body characteristics to environment which are found in a wide variety of creatures; e.g., Gloger's rule (1833) relates pigmentation to temperature and humidity. Indeed, the suggestion that ecological rules were applicable to man by Ridgeway in 1908 may well have stimulated the earliest studies of the effect of environment on morphological features. Experimental and theoretical genetics, besides clarifying the mechanism by which hereditary makeup varies, has shown that identical genotypes may develop into different phenotypes in different environments. During the postwar era, the detailed and intensive collection of objective data on a number of different animal species, so well summarized by Huxley (1942), Mayr (1963), and others, has allowed further clarification. The geographic variation observed in a species is an inevitable consequence of the geographic variation of its environment. The species has to adapt itself in different parts of its range to the local demands, so that every local population is under continuous selection pressure for maximal fitness in the area where it occurs. In the external environment, climate is one of the major groups of factors that may exert selection pressure on the phenotype. Climatic factors usually manifest themselves differently from those of habitat and from biotic factors in their effects on geographic variation. Climate tends to remain constant over long periods of time (by comparison with organism lifespan); consequently, selection pressures operate in the same direction generation after generation. Macroclimatic factors usually change rather slowly over wide areas, except where altitude is involved, whereas biotic and other nonclimatic habitat factors are often very local and irregular in their variation. Hence characters that vary with climate tend to show clinal variation, by contrast to the variation that is produced in response to many other types of environmental variation. Moreover, climatic gradients occur in parallel in different continents so that intercontinental comparisons can be made; and if populations long separated genetically show parallel morphological gradients, the case that they are associated with climate is strengthened.

Climatic factors can theoretically be expected to influence the form of any organism in the following ways:

1. Climatic factors may influence the genetic potential (the genotype) of the organism, determined by its position in the evolutionary sequence and its rate of change:
   - (a) directly, by promoting the selection of those morphological, physiological, mechanical, and other features which promote greater organic efficiency in a given environment;
   - (b) directly, by affecting cytological processes;

(c) indirectly, by influencing the size of population a given region will support, the effectiveness of mutation and selection being dependent on the size and breeding structure of the population;
(d) indirectly, by influencing the degree of accessibility of a population, and hence the degree of isolation in which it develops.

2. Climatic factors may influence the form itself of the organism (the phenotype) in its development from its genetic endowment:

(a) by direct modification of gene manifestation, induced by such factors as temperature or humidity;
(b) indirectly, by conditioning the amount and quality of those materials necessary for organic existence, growth, and development.

It therefore seems reasonable to suppose from general biological evidence that climatic relationships of body characters may well be demonstrable in man.

Part I of this book begins with an examination of the empirical evidence to see whether adult physique in humans shows any relationship to climate. Then the physiological basis for the relationships is explored, and the experimental evidence that morphological and ethnic differences affect response to climatic stress is reviewed. Part II considers the developmental aspects of the associations, the way in which they are brought about, by examining first the evidence on climatic associations of physique in childhood, and how physique is related to nutritional differences, and finally the relationship of growth processes to climate. In Part III, we explore climatic associations of features other than whole-body morphology for which evidence is beginning to accumulate, with particular attention to nose shape, the form of head and face, and skin pigmentation.

PART I

The Association with Climate of Adult Physique

Chapter 1

The Empirical Evidence

MATERIAL AND METHODS

To collect afresh data for an empirical examination of the climatic distribution of human physique would be a gigantic undertaking. Fortunately, buried in the literature of physical anthropology is an immense accumulation of objective quantitative and semiquantitative information relating to the appearance of the human body. This material was collected over the last century by those whose prime concern was with morphological variations from race to race rather than from place to place. That they tend to ignore environmental considerations, however, makes the data the more valuable, since there can be no bias for or against any climatic hypothesis.

On the other hand, there are disadvantages. The very characters that one would like to examine are not necessarily included. The material is unbalanced: of head measurements there is an abundance, of body measurements a sufficiency, of physiological measurements barely enough for exploratory analysis. Variations in technique of taking a particular measurement or of sampling procedure, which would be avoided in a prospective study, must be considered. Difficulties arise from the dispersal of the data throughout the literature with no comprehensive bibliography, from the nature and limitations of the data, from its sheer bulk. These can be overcome by adherence to somewhat arbitrary rules for rejection of questionable samples, e.g., excluding samples in which there was no differentiation of the sexes, excluding samples of fewer than 20 individuals for anthropometric characters, excluding individuals who were not yet adult or were senile. The absence of genetic data about populations is a special disadvantage, particularly when one is attempting to classify the samples into comparable groups as a preliminary to estimating the relative effects of race and environment. In what follows, the groupings used vary slightly from section to section, the object being to make the best use of the samples available for each.

The samples chosen for examination relate only to indigenous peoples. Europeans in extra-European areas and other recent migrants are excluded (except for occasional reference as a matter of interest, particularly in the physiological measures). It may be asked whether pre-European population movements in the more primitive areas, about which comparatively little is known, would not invalidate any analysis of climatic relationships. But in contrast to the vast spaces covered by recent European and European-associated movements, many other known migrations occurred only over distances involving relatively little climatic change—for example, the movement of the plateau Bantu into southern Africa, the movement outward from southeast Asia into Indonesia, the latitudinal migration of Arabs across northern Africa, and the movements across the Sudan grasslands. Second, where migration may have resulted in change in climatic environment for the group, the inclusion of relative newcomers to a particular habitat would tend to obscure any underlying environmental relationship, so that any relationship observed would be in fact closer than it appeared.

The method of analysis employed in this empirical review consists of first obtaining from the literature data for the given character on peoples throughout the world; second, screening the samples to maximize comparability; third, visual comparison of the geographical variations they show with climatic distributions; and fourth, statistical examination,[1] inquiring by correlation

---

[1]Perfect correlation between two characters, in which change in one produces an exactly proportional change in the other, gives the maximum value of the correlation coefficient (1). When the two characters are completely independent, the correlation coefficient is zero. The correlation coefficient is positive when change in one character produces change in the same direction in the other, and is negative when increase in one produces decrease in the other. Thus the correlation coefficient can take any value between -1 and +1, and so indicates the degree of association between the two variables. The linear regression coefficient measures the amount of change of the

and regression methods into the reality of associations so suggested, and then into the existence of responses at different levels among continental groups of man. The implications of the findings are then examined.

BASAL METABOLISM

Of the complex of variables of which climate itself consists, the climatic perception of the human body depends essentially on a combination of three: the heat (i.e., the dry bulb temperature), the humidity, and the windspeed. Warm-blooded mammals maintain their body temperature by producing more heat than is required and releasing the surplus to the environment, and a human is no exception. Heat is thus a variable of the environment and of human physiology, and human heat production is an obvious first choice for examination. Moreover, it is a physiological axiom that heat production varies with physique, so it may well provide a clue to physique variations with environment. Finally, heat production, measured as basal metabolism, is one of the few physiological measures that have been studied in non-European races.

Basal metabolism may be defined as the minimal level of heat production of the body when it is at rest (not asleep), is not engaged in heavy

---

[1] (cont.) dependent variable per unit change in the independent variable. In partial correlation, when there are more than two variables, the effect of one independent variable is examined while the others are held constant.

digestive processes (i.e., is in a postabsorptive condition), and is not undergoing marked mental activity or emotional stress. It is conventionally measured under ambient temperature conditions of 20°C. Basal metabolism represents the total energy utilized in maintaining those body states and processes necessary to animal life—body temperature, glandular activity, respiration, circulation, peristalsis, muscle tone, etc. It shows a normal curve of variation, as do many biological characters, from one individual to another. It may be measured directly with elaborate apparatus, or indirectly by calculation from the respiratory exchange of gases; the indirect methods are usually employed in the field. The amount of heat produced is computed usually from the amount of oxygen absorbed, rarely from the amount of carbon dioxide eliminated, or sometimes from both.

There are available for analysis 160 samples of basal metabolism in adults in different parts of the world, and it is possible to assign these to seven categories ranging from "very high" to "very low." The geographical distribution of these categories shows that "very high" basal metabolic rates occur only in Arctic regions, and "high" levels only in the temperate regions of central Chile, Kweichow, and the mountainous areas of central Java. "Fairly high" levels do not occur within the tropics except in the case of American Indian indigenous groups. "Medium" basal metabolic rates occur at all latitudes, but the majority of samples in tropical regions are in the lower categories, while "very low" levels do not occur outside the tropics. In India, European samples give results higher than those of indigenous locals, while in America, American Indian peoples give higher results than do Europeans. In brief, a distinct pattern related to environmental temperature is suggested, and at the same time there appear to be differences among continental groups.

Statistical examination of the association between daily basal calorie production of adult indigenous male samples and mean annual temperature shows a highly significant correlation ($r = -.736$), and female samples give a confirmatory result. Significant differences also appear among the great varieties or continental groups of men, the American Indian being consistently high. Within each such continental group the negative correlation with mean annual temperature of the habitat remains, and is sufficient to account for many of what have been taken to be race differences in basal metabolism (Fig. 1). The slopes of the regressions remain very similar if the high-protein diet samples are omitted, and the regressions cannot therefore be attributed to different amounts of animal protein in the diet. Over 81 percent of the variation in mean daily basal calorie production is attributable to variations in mean annual temperature, stature, and weight, and each of these three factors exerts an independent effect.

When basal metabolism is expressed per unit weight and per unit surface area, similar associations with environmental temperature appear. But with increasing environmental temperature the rate of decrease of heat production per unit surface area is greater than the corresponding rate of decrease in heat production per unit weight, after allowance has been made for change of unit. Further, this difference in rate remains after elimination of variation between continental groups. Therefore, the function surface area per unit weight must increase with increased temperature,

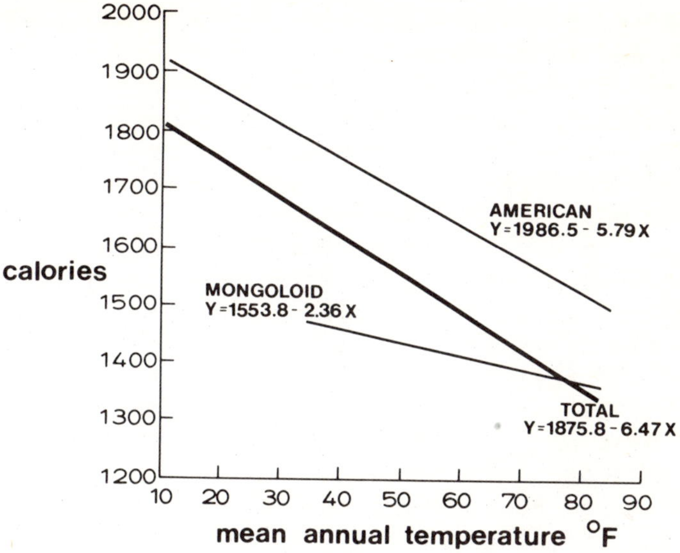

FIG. 1
Total regression of daily basal calorie production on mean temperature and some of its component intravarietal regressions.

and the function weight per unit surface area must decrease with increased temperature. In other words, since surface area per unit weight is a measure of linearity of form, the human body tends toward sphericity with decreased temperature and toward linearity with increased temperature.

There are intervarietal differences (i.e., between continental groups of man) in heat production after allowance for the influence of body size and environment, but intravarietal differences largely reflect variations in climate. But more important for the study of the morphology of the human body is the fact that from these relationships can be deduced the tendency for linearity of form to increase with increased temperatures and to decrease with lower temperatures, the relationship of surface area to weight becoming respectively larger and smaller (Roberts, 1952b). Thus a mechanism is provided by which variations in body form in response to environmental demands can be understood. The hypothesis may be briefly stated: In order to facilitate the balance of human body heat exchange, under those conditions in which heat loss is more difficult the amount of body tissue is reduced, so that less heat is produced and the ratio of surface area to weight (i.e., the ratio of potential heat loss to potential heat production) is increased. The next two sections are devoted to an examination of this hypothesis. Body weight is chosen as the measure of mass; for the surface area/weight ratio, in the absence of direct measures of surface area, metrical characterization of limbs and trunk is preferred.

BODY WEIGHT

Body weight may be taken as a measure of body size. The geographical distribution of mean body weight in 116 comparable samples of indigenous male groups shows that the lowest weights are not found outside the tropics and the highest weights are not found in latitudes lower than 30°. Comparison with the isotherms of mean annual temperature suggests a general temperature relationship; for example, very few of the samples in the weight categories heavier than medium fall on the warmer side of the isotherm for a mean annual temperature of 50°F, and only one sample lighter than medium falls on the cooler side of the isotherm. The correspondence is, of course, not perfect, a few samples showing considerable deviation, to be attributed not only to sampling error but also to the operation of factors such as nutrition and stature. The effect of stature can be partly overcome by examining weight per unit stature, and the same pattern appears.

Statistical evaluation of the suggested temperature relationship gives for male adults a correlation and regression coefficient of $r = -.600$, $b = -.305$ kg/°F. Now weight obviously varies with stature, and in the present series there is a significant association between stature and temperature ($r = -.351$) and between stature and weight ($r = +.734$). However, partial correlation shows that the correlation between weight and temperature exclusive of stature remains highly significant ($r = -.538$), and the correlation between stature and temperature disappears. There is a multiple correlation coefficient $R = .820$, indicating the importance of the joint effect of stature and temperature on the weight of a sample. Similar calculation using the cube of stature gives the multiple correlation coefficient $R = .823$, making very little difference. A shorter series of female samples gives a weight/temperature partial correlation coefficient $r = -.704$.

If we examine the differences between continental groups of man, these remain highly significant after allowance is made for the association with temperature and stature, the mean adjusted for their effects being lowest among groups thought to have occupied a tropical habitat for a considerable time, Australian and South Asian aboriginal and African populations. But also within these continental groups, the temperature relationship remains highly significant; the partial correlation coefficient between weight and

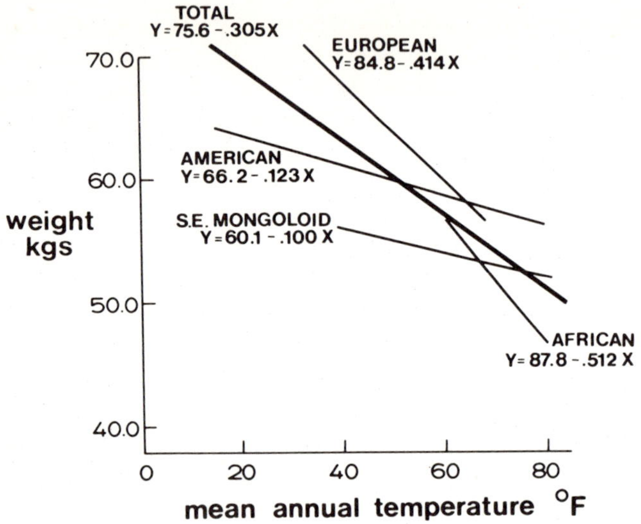

FIG. 2
Total regression of weight on mean temperature and some of its component intravarietal regressions.

temperature exclusive of stature is r = -.362 (Fig. 2). The within-group correlation coefficient is R = .753 (Roberts, 1953; Newman, 1960).

Thus the negative relationship between the mean weight of human groups and mean temperature, postulated from the basal metabolism study, is confirmed in this larger series of data. The depression of mean weights in tropical areas and their elevation in cooler areas is accounted for in part by the affinities of the samples with particular continental groups, in part by a tendency within such groups. Variation of stature, marked though it be, scarcely affects the temperature relationship. The data are not sufficient to show whether the relationship between weight and temperature is direct, constituting an immediate response to temperature, or indirect, acting through inheritance or through the intermediaries of type and amount of food supply, habits, etc. If indeed lower weights are advantageous in warmer areas, then so long as that condition is attained, it appears immaterial whether it is attained directly or indirectly. In any case, all responsibility should not be attributed to nutrition, for this would not explain features such as the European correlation, the high non-European cold results, the low figures of adequately fed tropical dwellers. The lowest weights do not appear in areas of greatest population pressure, nor do the highest weights always occur where living conditions are easiest. Further, if nutrition were the cause, one might expect a certain nutritional influence during the growing period, affecting stature, for example, so that elimination of the stature factor should markedly reduce the correlation between weight and temperature; this clearly does not occur.

BODY SHAPE

The Lower Limb

As a character indicative of the relative lengths of trunk and lower limbs, sitting height occurs more frequently in the literature than any other measurement from which leg length may be obtained. Sitting height represents the vertical length of the head, neck, and trunk; thus when expressed as a percentage of stature (the relative sitting height) it gives a relative measure of the length of the lower limb. There is considerable variation among techniques recommended for taking the measurement, which has to be considered in interpreting any comparison.

Examination of a series of 300 samples of males from different populations throughout the world shows a marked tendency toward regional grouping of relative sitting height mean values. Although medium values occur at all latitudes, in Africa the bulk of the lowest values occurs in the hottest regions, in a broad sweep across the northern tropical grasslands, with low to medium values tending to occupy forest (except for pygmy samples) and less hot areas, and medium levels occurring chiefly in the Mediterranean area. Asia's lowest means, generally not so low as those in Africa, appear to congregate in the warmer south, with high values occurring in the cooler mountain regions and areas of colder climate, while the highest values are found only in the very cold territories. American samples with low means occur only in the hot desert areas and in the lower hotter areas of Central America, while high values occur at higher altitudes and latitudes and very high values only in the Arctic and Subarctic and in a single Andean sample. European samples, all at medium to fairly high values, show little distribution pattern, nor do the few samples from the Pacific area. Again, a relationship with mean temperature is suggested.

In analysis of a series of 285 samples of males, the correlation and regression coefficients of actual sitting height and mean annual temperature ($r = -.428$, $b = -.797$) demonstrate a highly significant linear association, almost one-fifth of the total variance of mean actual sitting height being ascribable to variation in mean temperature. For relative sitting height, there is a

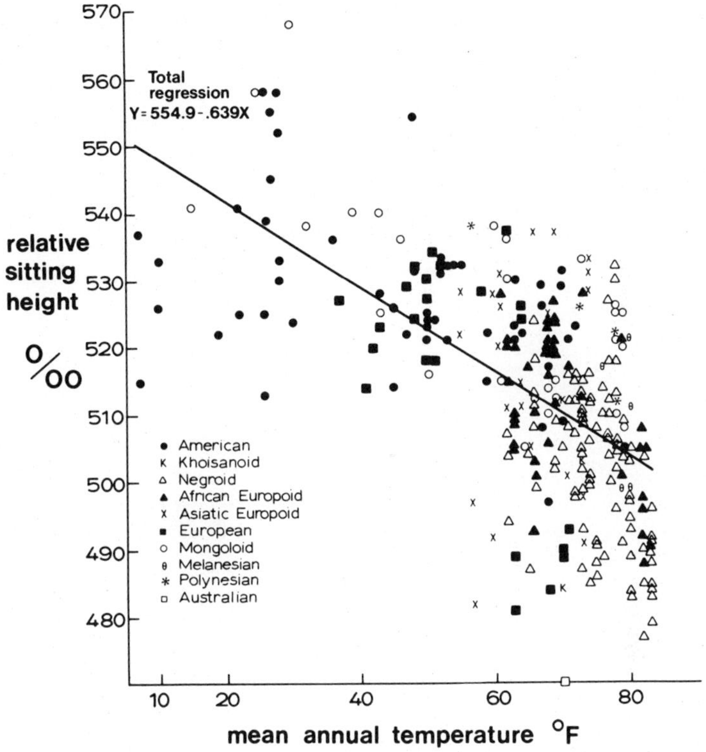

FIG. 3
Relative sitting height and mean annual temperature, for male indigenous groups.

closer and again highly significant linear association with mean annual temperature (r = -.619, b = -.639). These values indicate that nearly two-fifths of the total variance is ascribable to mean temperature, and therefore that body proportions are more closely related to temperature than is absolute size (Fig. 3). This marked relationship also appears in a series of female samples, again the values of r = -.654, b = -.700 being highly significant.

The accumulated series of samples included representatives of many races, drawn from each of the great varieties or continental groups of man. When the samples are classified into ten such groups, covariance analysis indicates that after allowance for the differences between these groups, there remains a highly significant linear relationship within such groups, with within-variety values of r = -.386, b = -.404. The difference in the values of the total and within-variety coefficients suggests that, although parallel temperature relationships exist among races within each variety, each variety is itself adjusted to a particular habitat range (Fig. 4). Indeed, comparison of the adjusted means of varieties, after allowance has been made for the temperature association, shows highly significant intrinsic differences among varieties. Lower values than the overall mean occur in Australian, Negroid, Khoisan, and Melanesid groups, i.e., groups which are generally thought to have occupied a tropical habitat for a considerable period. Higher values in tropical people are found in those who may well have entered hot habitats relatively recently. Although differences among the regression coefficients are not significant, it is interesting that of the four groups which show individual temperature regressions significant at 0.1 percent, the lowest value of b occurs in the American series. Female samples also show highly significant intravarietal temperature relationships with r = -.403, b = -.276, and again highly significant differences between adjusted means.

Statistical investigation thus confirms the relationship of relative sitting height with

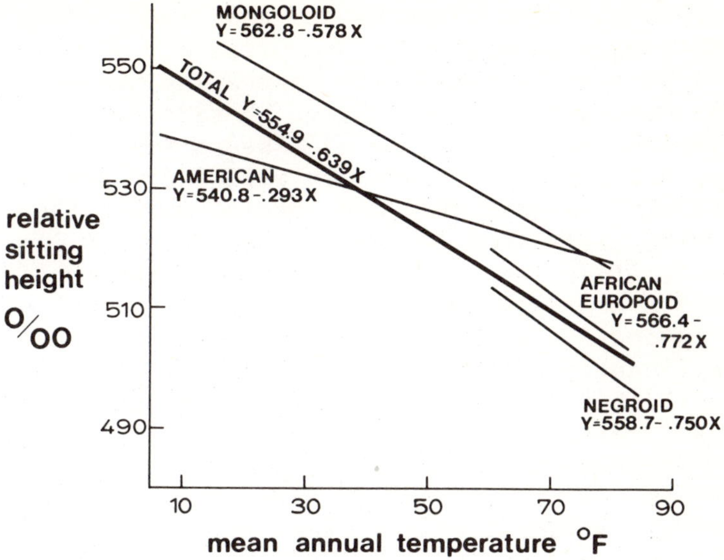

FIG. 4
Total regression of relative sitting height on mean temperature with some of its compound regressions.

temperature suggested by the general geographical distribution pattern, while the agreement between the fairly independent male and female series provides a further check. It is likely that the values of the correlation coefficients will be rather low on account of the additional variability due to the material and method, but for the moment they may be regarded as indicative of the close link between climate and the relative length of the lower limbs.

Also available in the literature are details of several other features of the lower limb. A similar statistical analysis of these showed that symphysial height, navel height, height of anterior superior iliac spine, and trochanteric height, each expressed in proportion to stature, all appear to increase with increasing temperature. Moreover, the lower limb, besides increasing in length relative to stature, seems to change its shape and become more slender with increasing temperature. A significant decrease occurs in the relative calf circumference and an increase in relative foot length. There is also a tendency toward a decrease in relative thigh circumference and in foot index. Relative calf length increases significantly with temperature, and so does the calf/thigh index, in which calf length is expressed relative to thigh length.

It seems that the distal portion of the limb increases in linearity more markedly than does the proximal, giving the characteristic elongation of the lower segment so frequently noted in African reports, but also characteristic of dwellers in hot climates of other continental groups.

The Upper Limb

The most frequent upper limb measurement in the literature is span, expressed either directly or as a percentage of stature (relative span). This is not a good measurement for purposes of the present inquiry, since it is a composite dimension comprising arm length and shoulder breadth, which are linear and transverse components respectively, and it appears that these components may be contrarily associated with climate. Again, there are variations in technique to be taken into account.

Geographical distribution suggests that in Africa values for relative span seem to decrease from the hot grassland regions, through the areas of grassland where the temperature is modified by altitude, equatorial forests and hot deserts, to the cooler Mediterranean and Abyssinian mountain regions. The pattern is not quite so clear as in the case of relative sitting height, but it does exist. Similar gradients are suggested in other continents.

For 372 male samples the correlation coefficient of mean actual span with temperature is r = +.235, which is highly significant, of mean span with mean stature r = +.898, while no significant relationship appears between stature and mean annual temperature. Thus span is largely conditioned by stature, but nevertheless an association with temperature is still detectable. Partial correlation shows the independence of the temperature effect. Though the partial correlation coefficient of span with stature, r = +.918, shows the importance of the stature relationship, nonetheless span and mean temperature have a partial correlation coefficient of r = +.484. It thus appears that while the span of any particular human group is chiefly determined by its stature, the temperature of its habitat exerts a modifying influence. The same point can be demonstrated by the correlation and regression coefficients of mean relative span and mean annual temperature, r = +.470, b = +.703, indicating that more than one-fifth of the total variance in mean relative span is attributable to environmental temperature (Fig. 5). Multiple covariance analysis indicates highly significant differences in span between continental groups after allowance has been made for the effect of temperature and stature. Within groups the association between span and temperature after allowance for the effect of stature remains highly significant.

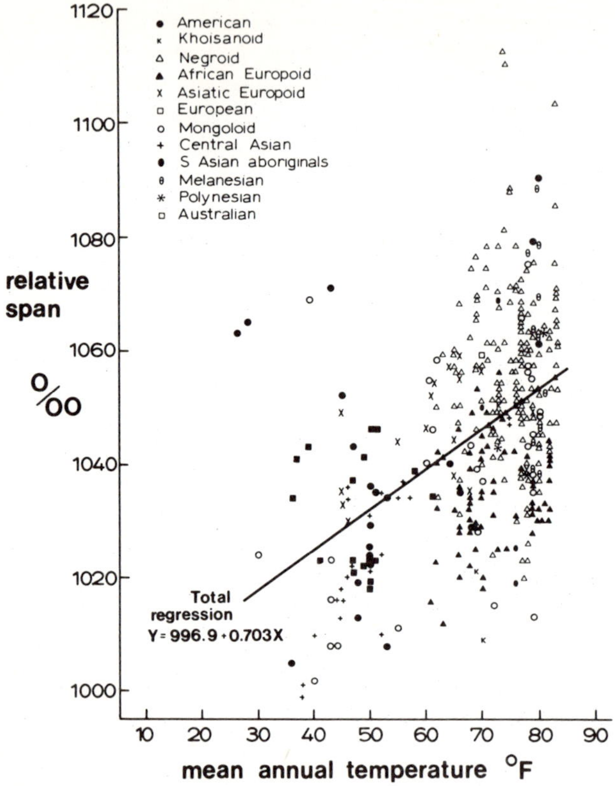

FIG. 5
Relative span and mean annual temperature, male indigenous groups.

As regards other features of the upper limb, relative arm length tends to increase with increased temperature. This appears to be due to tendencies to elongation with increased temperature in each of the three segments of the arm (upper arm, forearm, and hand), all relative to stature. However, the antibrachial index (forearm length expressed as a percentage of upper arm length) also increases markedly with temperature, whence it appears that the segments are not equally elongated but that greater increase occurs in the lower arm than in the upper. Now the forearm is generally more slender (absolutely) than the upper arm, so that less increase in volume would be produced by a given increase in length. In other words, greater modification of the mass/surface area ratio is effected by increase in the length of the lower, than of the upper, arm. Hand breadth, upper arm circumference, and the hand index show no significant relationship, partly because the measurements themselves allow considerable technical variability and they are all affected by use and daily habits.

The Trunk

Of the measured characters of the trunk, chest girth is the dimension most frequently given in the literature, probably because it is so easy to measure. But this ease of measurement makes for great variation in the method of measurement, and differences of up to 80 mm may occur in the recorded chest girth of a given individual according to which technique of measurement is used.

Nevertheless a geographical pattern emerges from simple examination of the data. Smaller chest girths are concentrated in warmer regions, greater values in cooler regions. This distribution is not due to technique, because it appears when results obtained by the same method are compared. Statistical examination of mean chest girth against mean annual temperature in 133 male samples gives correlation and regression coefficients of $r = -.575$, $b = -.155$, which are highly significant (Fig. 6).

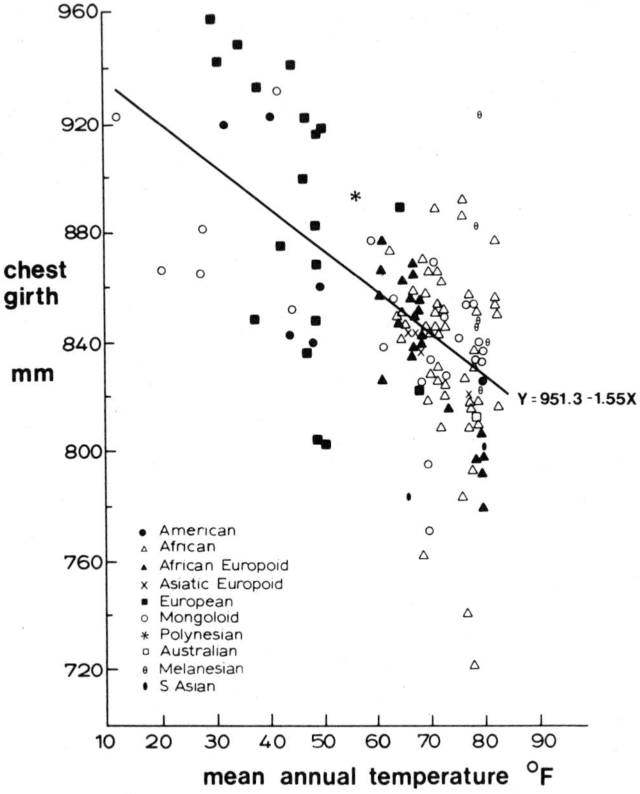

FIG. 6
Chest girth and mean annual temperature.

Much of the variability about the regression line may be explained by the differences in technique, so that the actual correlation is probably considerably closer. A further factor contributing to the variability is stature, with which chest circumference is correlated. However, the temperature relationship remains after the effects of stature are considered, the partial correlation of chest girth and temperature being r = -.585; thus the relationships of chest circumference with climate and with stature are independent of each other.

Of other trunk features that have been recorded occasionally, both relative bicristal breadth and relative biacromial breadth show highly significant associations with mean annual temperature, giving correlation coefficients of r = -.823 and r = -.577 respectively. In cooler areas, then, both the pelvis and the shoulders appear to be broader than in warm regions.

THE CLIMATIC RELATIONSHIP OF PHYSIQUE AND THE ZOOLOGICAL RULES

The zoological rules relating body size and proportions to environment state: "within a polytypic warm-blooded species, the body size of the subspecies usually increases with decreasing mean temperature of its habitat" (Bergmann's rule, 1847) and "in warm-blooded species, the relative size of exposed portions of the body decreases with decrease of mean temperature" (Allen's rule, 1877). These rules have their foundation in physical laws. The larger the surface area of a body, the greater the loss or gain of heat by convection and radiation, and the greater the area over which evaporation can occur. For humans the last law is particularly important at high temperatures, where heat loss is essential and sweating is the main outlet for it (see pp. 14-15). Second, the amount of heat produced in working depends largely on the body weight of an individual. The heat produced per kilogram of weight in a given task tends to remain constant whether the individual is large or small. But the heat produced will not be constant per unit surface area, since increase in size of a body does not produce the same effect on body weight and on body surface area. For a given change in a linear dimension, the surface area changes as the square, and the body weight as the cube. Hence a large person of the same proportions as a small person will have, per unit of heat produced, a smaller surface available for its dissipation. Thus he or she will need to lose more heat per unit area. Third, for a cylindrical body, the amount of heat lost per unit area for a given environmental difference increases rapidly as the diameter of the cylinder is reduced below a threshold level. Hence an individual with slender limbs is by physical laws able to lose relatively more heat than a person with sturdy limbs.

The discussion of basal metabolism appears to confirm the application to man of the functional mechanism embodied in these rules; namely, in order to facilitate the balance of human body heat exchange, under those conditions in which heat loss is more difficult, less heat is produced (through reduction in the amount of body tissue) and the ratio of surface area to weight (i.e., the ratio of potential heat loss to potential heat production) is increased. Analysis of weight shows the general applicability of Bergmann's rule to humans, and also that it needs some modification. Clearer definition of "body size" is necessary. Defined by reference to linear dimensions (e.g., stature), it is not applicable, whereas defined in units of mass (e.g., body weight), it is not only applicable but needs restatement to incorporate, with the postulated decrease in size among subspecies (i.e., continental groups) within the species, similar diminution within the subspecies.

There are disadvantages to using body weight. It is a gross, complex character, combining indistinguishably measures of the amount of metabolically active tissue, of insulating tissue (e.g., subcutaneous fat), and of other metabolically inactive tissue. Thus the relationship demonstrated may indicate in warm areas an actual reduction in the amount of heat-producing tissue or a decrease in the amount of insulating tissue or both. More detailed analyses of body composition are required to clarify the demonstrated relationship. Moreover, it is a labile character, varying over short periods with food consumption and health. The climatic correlation may thus involve nutritional differences, or perhaps the residue of nutritional differentials during growth rather than genetic adaptation; the relationship of climate and weight observed in white Americans cannot be attributed to long genetic selection, and Newman and Munro (1955) inclined to increase of appetite in cold conditions as the explanation. Certainly, in indigenous New World peoples the correlation of mean weight with mean temperature of the coldest month ($r = -.729$) is higher than that with mean annual temperature. However, for one function, body weight can be considered as an entity. It represents the mass to be carried in any body movement, and so the amount of heat produced by a man doing physical work is closely correlated with his weight. Hence the relationship here demonstrated may indicate a further advantage in hot climates in that the energy expended in any activity requiring movement of the whole body will be less if the amount of weight to be carried is less; this again may be of appreciable value in hot areas. There are also other sources of advantage. The specific heat of adipose tissue, of low water content, is only half that of the fat-free body mass; adipose tissue therefore has greater insulative value, so that a given heat load per kilogram of body weight will cause greater temperature elevation in fat individuals. Again, the cardiopulmonary system of obese individuals is inferior to that of lean individuals, and people with compromised cardiovascular function are excessively intolerant of heat stress. These mechanisms may well contribute something to the weight/climate relationship.

The sections on body shape, insofar as the "exposed portions" were restricted to the limbs, indicate that Allen's rule requires similar redefinition, to separate "relative size" into its components of linearity and bulk, to envisage intrasubspecific variation, and to provide for the fact that the development of one region or of one characteristic may obviate modifications of any other. Alternatives are therefore possible. The contrasting physiques of the squat pygmy and the tall linear Nilotes represent alternative ways of maximizing surface area relative to weight and hence alternative advantageous forms in hot conditions. However, all relationships found (summarized in the table on pages 94-95) are in accordance with the general principle (of which Bergmann's and Allen's rules are but particular aspects) that body surface area relative to mass in warm-blooded creatures tends to increase with increased temperature. Hence this general principle seems to be applicable to humans.

The method employed here is crude, and hence the values of the correlation coefficients may be too low and the actual degree of relationship between the body feature and the environment may be considerably closer than the coefficients indicate. It is possible that curvilinear expressions would show yet a closer relationship. Indeed, interpretation of the correlation coefficients should be guided by the direction in which they point rather than by their actual numerical values. The various body characters considered are by no means independent and represent the outward and visible signs of underlying biological factors which, were they extracted, might show closer correlation with climate. In addition to the variability due to sampling, technique, and nutritional and other factors innate in the material, it is necessary to consider the unavoidable error due to imprecise sample location, to the necessity to estimate temperature, to as yet unconsidered variables of the climatic environment, singly or together. Mean annual temperature itself is used only to demonstrate the presence of associations, and examination of some more meaningful physiological value such as effective temperature or the P4SR index is the next requirement in the analysis. Certainly it conceals the range of conditions at which human groups are to be found; there is a difference of 140°F in mean temperature of coldest months between the warmest and coldest human permanent settlements (Insalah 80°F, Verkhoyansk -60°F), and again these are averages, masking yet greater actual extremes. But mean annual temperature is not meaningless, and though it is not a measure of climatic stress, it probably reflects the frequency of occasions of stress of varying intensity throughout the year and it is a good conservative general measure of climate difference. And only one period of very extreme conditions per generation may be required for intense selection to take place.

From these findings the following generalizations may be made:

1. In indigenous peoples there is a marked relationship between the morphology of the body and the climate of the region inhabited; this applies within continental groups as well as overall. The values of the correlation coefficients are sufficiently high to need explanation.

2. In those series which are numerically adequate there appear to be differences among the major varieties of humans even after the influence of temperature has been taken into account.

3. The pattern of the relationship is made up of a series of positive correlations with temperature in those characters causing increased linearity of body form, and a series of negative correlations in those characters causing increased mass at the expense of linearity. Variation in every character is such as to increase the body surface area relative to body mass with greater temperature. The elongation of the distal portions of the limbs relative to the proximal, as indicated by the calf/thigh index and the antibrachial index, represents adjustment of those regions of the body where, by reason of their smaller diameters, an increase of given length produces the maximum surface area increase and the minimum mass increase.

4. These variations may be described by Bergmann's rule and Allen's rule (with minor modification), formulated to describe similar variations long known in the animal world, whose functional basis concerns heat production and dissipation, the facilitation of the maintenance of the body heat balance. Under extreme conditions of climatic

stress, such adjustment is likely to be of considerable importance, and hence the basis may lie in genetic adaptation through long-period selection.

Chapter 2

The Physiological Basis

The results so far suggest that humans possessing different morphologies may respond differently under conditions of climatic stress. The human body functions efficiently only within a very narrow range of internal body temperature. This is maintained by a sensitive and complex thermostatic mechanism which responds to changes in the body heat load, itself affected by the external conditions (Hardy, 1961; Collins and Weiner, 1968; Benzinger, 1969; Kerslake, 1972). The physiological changes that occur under extreme climatic conditions may perhaps indicate the functional reasons for the morphological association.

When a person enters a hot environment, a number of physiological changes occur: the skin temperature rises, the surface blood vessels dilate, there is a slight increase in blood and plasma volume, and more blood flows through the skin, carrying more heat to the periphery to be dissipated. The blood pressure drops, with the result that the heart output increases to maintain blood pressure, and the pulse rate increases. When the ambient temperature rises above about 80°F, less heat is lost by radiation and convection and one begins to lose more and more heat by evaporation of sweat, until, when the ambient temperature rises above the body temperature, the body is actually gaining heat by radiation and convection and the only means of heat loss is by sweating. As the environment becomes more severe, or if the amount of heat one produces internally through increased work load markedly increases, there is first discomfort, then distress, and eventually there comes a point when the sweat glands break down, one can only store heat, and collapse occurs. Under extreme conditions, particularly where evaporative loss is difficult by reason of high humidity, the importance of a slight increase in area over which sweating occurs is clear. There may also be other unpleasant effects. Hyperventilation may so markedly increase the loss of CO_2 through the lungs, besides CO_2 lost through the sweat, that there may be an increase in blood pH; if this change is rapid, tetany may occur. The peripheral dilation and sweating pose a secondary threat to cardiovascular function through inadequate blood volume and dehydration.

When a person is exposed to cold, the body loses heat primarily by conduction, convection, and radiation, the rate depending on the difference between body surface and environmental temperature and the amount depending on the rate and the extent of body surface. Heat is also lost by evaporation from the lungs and by insensible perspiration. The skin temperature drops, and often vasoconstriction diminishes the conductivity of the peripheral tissues. But the limit of its effect is soon reached, after which skin cooling continues to occur. Also, cycles of dilatation subsequently occur, apparently preventing too great cold in the surface tissues. Heat continues to flow from the deep tissues to the surface, so that the deep body temperature begins to fall. To maintain thermal equilibrium one augments heat production. The time taken to initiate the response depends on the environmental temperature: less than two minutes on exposure to 3°C, less than seven minutes at 10°C; some increase in metabolism occurs at exposure to any environment cooler than 26-28°C, the "critical temperature." This increase involves visceral and muscular tissue metabolism, voluntary movements (exercise), and the reflex onset of shivering. The metabolic response, which is proving extremely complex physiologically, is triggered both by an abundant inflow of information from the peripheral cold receptors, which causes shivering and via the hypothalamus activates the thyroid, and by the drop in core temperature itself. Indeed, if the deep body temperature continues to drop, coma, muscular rigidity, and death ensue. In cold conditions the diminished surface-area-to-mass ratio clearly is important.

The physiological adjustments involved in these processes are sufficient to explain many of the so-called "racial" differences observed in measures of function in indigenous human groups (Roberts, 1952a). Respiration rate and pulse rate have positive correlations with mean environmental temperature, and blood pressure has a negative correlation. The increase in respiration rate with temperature may be associated with greater evaporation from the lungs per unit time, a by no means negligible component of the total heat loss of the body. Similarly associated with the maintenance of heat balance, the increased pulse rate with temperature most likely reflects an increased heart output to allow a greater quantity of blood, per unit time, to be sent to the periphery for cooling in its passage through the surface vessels. The blood pressure/temperature relationship also seems connected with body temperature regulation; in cooler climates, a greater degree of vasoconstriction increases the peripheral resistance; blood pressure is therefore higher, skin temperature is lower, and loss of heat by radiation to the surrounding air less.

Unfortunately, such simple rationalizations must be qualified. The amount of heat lost from a body depends not only on its surface area, but also on the insulating and radiating quality of that surface. In vitro the thermal conductivity of fat is only about half that of muscle, so differences in the thickness of subcutaneous fat must be considered, though of course humans do not have the massive layers of insulating fat, and certainly not the fur, that are so useful in Arctic animal species. However, available data to date suggest that there are gradients in skin fold thickness parallel to those demonstrated for the other anthropometric characters. Radiation and absorption also vary with skin color. The skin temperature of pigmented skin rises more rapidly than white when exposed to heat (see p. 83).

Another complication is that man becomes habituated to near-extreme conditions by the process of acclimatization, defined as that adaptive process which results in the reduction in the physiological strain produced by environmental stress (Bass et al., 1955). As one becomes acclimatized to heat, the heart rate slows and circulatory strain is reduced by changes in the fluid distribution, e.g., isotonic expansion of the interstitial fluid. The internal temperature reaches a new lower level, and so does the skin temperature. The body is no longer concerned with pushing as much heat as possible to the periphery, but instead it begins to sweat earlier and increases the rate of sweating at a given body temperature; the sweat glands become more sensitive to stimulation. To avoid the serious drain of water and salt from the body, a new balance is established and the sweat becomes more dilute. There is a concomitant improvement in one's ability to perform muscular work, a noticeable reduction in the feeling of discomfort, and an improvement in the performance of intricate tasks. This process has been shown to be essentially the same in all major races. Cold tolerance can also be developed over a period. There is a change in the pattern of dietary desires toward increased fat consumption. A higher level of metabolism allows higher deep body and skin temperatures to be maintained. There is improved blood flow through the extremities, and tactile sensitivity and manipulability at a given temperature improves. There is less discomfort. The complexity of the process implicates conditioning of the central nervous system, the sympathetic nervous system, and the endocrine glands.

The physical facts underlying Bergmann's and Allen's rules can now be put into physiological perspective. Apart from convection and radiation differences, a smaller individual in a given heat stress will have to sweat less than a large individual; of two people of the same size, the more slender person will need to sweat less than the fatter. Therefore, there will be a greater load on the sweat glands of the larger or stockier subject, while the smaller, more slender individual will be less pressured. This argument is quite independent of the amount of insulating fat. The converse obviously holds in conditions of cold stress. Interesting support for this argument comes from examination of the physique of champions in sports of different types, comparing, for instance, the lighter physique of the long-distance runner and the rounded, insulated physique of the long-distance swimmer, where the problems are respectively not to accumulate and to conserve heat (Carter, 1970).

EXPERIMENTAL EVIDENCE OF MORPHOLOGICAL EFFECTS ON CLIMATIC STRESS TOLERANCE

Physiological considerations suggest that morphological variations may be of significance in survival under climatic extremes. The experimental evidence is not extensive. It is a curious commentary on the large amount of experimental work undertaken that, initially, the factors of morphological and ethnic variation received so little attention. Yet many investigators commented on the surprising differences found between the reactions of individuals exposed to the same carefully controlled climatic environment. Today an appreciable body of literature is available (e.g., Adolph, 1947; Baker, 1958; Barnicot, 1959; MacPherson, 1960; Ladell, 1964; Carlson and Hsieh, 1965; Weiner, 1967).

Perhaps the most striking piece of evidence is that during the 1939-1945 war during intensive courses of desert training in the southwest of the United States, a number of soldiers collapsed and died (Schickele, 1947). For 157 such deaths the height and weight data of the unfortunate individuals were compared to those for a random sample of 100,000 soldiers. It appeared that the greater the man's weight relative to his stature, the greater his susceptibility to fatality. Men 5 to 14 lb above normal weight for their height were four times more susceptible than males of average weight, while males of average weight were four times more susceptible than those who were more than 16 lb below average for their height and age. It is unfortunately not clear that the cause of death in these cases was fatal heat collapse rather than some other condition, but this evidence is suggestive.

Under experimental conditions, the classic comparison was that by Robinson of a heavy man and a light man working at comparable rates under hot conditions (32°C, 70 percent humidity); it was found that the heavy man was unable to complete the task on account of approaching heat exhaustion, having accumulated heat steadily throughout the work. It is well established that sweat output is affected by body weight ($r = +.75$), sweat increasing at the rate of about 27.5 g/kg. However, correcting sweat output on the basis of weight reduced the observed differences between subjects while working, but in resting subjects a better reduction of individual differences was achieved on the basis of estimated surface area ($r = +.65$, sweat increases at the rate of 17 g/cm^2 surface). This suggests that the weight relationship merely indicates that a heavier subject has to do more work in moving his body and that it is sheer size that is mainly disadvantageous in heat stress. This is confirmed by a small investigation on four subjects which showed that heavier subjects had a higher rate of heat production per unit sweat loss in grams per hour, that their skin temperature when working did not drop as far as did that of lighter subjects, and that their sweat loss per unit weight was lower at a given walking speed; differences both in heat production and sweat loss between individuals remained after the differences in body weight were taken into account. Again, the heavier persons are at a disadvantage but their size is not the sole consideration.

There are several studies which mention linearity of form or composition of the body (Baker, 1955; Buskirk et al., 1955). When groups of men were subjected to desert heat stress, sweat loss and rectal temperature were higher in the fatter men, suggesting that under desert heat the fatter the man is, the lower his expected heat stress resistance; the amount of nonfat tissue in the body, however, also influenced sweat loss positively while the greater the degree of attenuation of the body, the lower the sweat loss. In other words, individuals who have a small amount of body fat and greater linearity can probably withstand heat stress longer than can those of contrasting physique. This conclusion also applies to women, for obese women showed greater heat strain than leaner women, as measured by rectal and skin temperatures, heart rate, and sweat rate, but the difference between the groups diminished at higher heat stress levels (Bar Or et al., 1969). In a recent direct comparison of cooling of seven obese women (fat 29-41% of body composition) and three lean (21-24% of fat) in a 20°C water bath (Kollias et al., 1974), the rate and magnitude of increase in heat production, and the decrease in core temperature, were significantly greater for the lean (with a mean surface area body weight ratio 2:32). It also applies in different racial groups. For example, responses to two levels of metabolic activity in heat over a 12-day period were measured in Shibipo and Mestizo South American men, and it was found that smaller size was associated with lower final temperature, lower final pulse rate, and lower sweat loss (Baker, 1966). Similarly, in Hanna's (1970a) comparison of indigenous Papago and nonindigenous European men subject to heat stress in the Sonoran desert, the apparent difference in sweat rate largely disappeared when the differences in body surface area were taken into account.

Individuals with much subcutaneous fat and with a stocky build can sit nude for longer periods in a cool environment with less loss of heat from the body and less metabolic disturbance. The thickness of the subcutaneous fat of the chest and abdomen is directly related to the amount of insulation it confers, and individuals with much subcutaneous fat are capable of tolerating lower skin temperatures with less discomfort than leaner control subjects. A highly significant correlation was shown between the amount of body fat and the drop in rectal temperature in men exposed for four hours to cold conditions. In cool conditions also (59°F), when almost nude men were physically inactive, fat men maintained higher rectal temperatures but had lower skin temperatures than thin men; when subcutaneous fat was held constant, less linear individuals maintained higher core temperatures at lower levels of heat production; thick subcutaneous fat provided more efficient insulation, as indicated by less shivering, lower skin temperature, and lower oxygen consumption. Over a long period of residence in Antarctica thin men increased in weight and skinfold thickness, and with this their metabolic and body temperature responses changed in the direction of those of fat men, who showed no change in weight or temperature response (Wyndham and Loots, 1969).

One of the most consistent morphological differences is that which occurs between the sexes, and of particular relevance is the difference in the amount and distribution of subcutaneous fat. Little et al. (1973) found that Quechua Indian females showed a generalized reaction to hand immersion in cold water that was different from the reactions of both Indian and white males. The females maintained warmer skin temperatures during immersion, and they recovered more rapidly thereafter. Moreover, the females rewarmed earlier with higher amplitude cycles, responded more actively during reactive hyperaemia, and showed lower systolic cold pressor and heart rate changes during the immersion. Such sex differences have been generally reported: Hanna (1970b) found that Quechua Indian women, during mild cold stress, maintained finger temperatures that were substantially warmer than men; Japanese and Ainu females showed greater cold tolerance of the fingers than did men during ice-water immersion (Kondo, 1969; Suzuki, 1969), and the females showed consistently low systolic cold pressor responses as well (Itoh et al., 1968). In the Arctic, Eskimos in Greenland and Canada, and Athabaskan Indians all show higher values for heat output from the hands of females (Lund-Larsen et al., 1970; Krog and Wika, 1971-2; and Williams et al., 1969). The mechanisms by which the sex difference is brought about are not clear, but it is possible that the females react to hand immersion in cold water by a lower level of stimulation from the vasomotor center of the brain (to account for the reduced systolic cold pressor and heart rate increases), perhaps by decreased peripheral sensitivity to cold, but certainly surface vascularization shows a greater dilating capacity in females (shown by the raised cold-induced vasodilatation, reactive hyperaemia, and rapidity of recovery). But whatever the mechanism, the differences are compatible with the morphological differences.

In summary, the experimental evidence is generally consistent and indicates the importance of morphological differences in climatic stress and appears to support the generalizations from the empirical analysis.

ETHNIC DIFFERENCES IN RESPONSE TO CLIMATIC STRESS

Heat

There is relatively little experimental evidence on the question of racial differences in climatic stress tolerance. In a comparison of six Europeans and three South Asian subjects in Singapore exposed to hot-wet and hot-dry environments before and after a 14-day acclimatization period, when due allowance is made for differences in size, the reactions of both groups were generally similar though in both groups the sweat output increased with acclimatization (Adam et al., 1953). However, the Asians always lost less sweat, both absolutely and in relation to their body size, than the Europeans (Asians 16.8 g/kg, Europeans 26.8 g/kg at the beginning; Asians 21.0 g/kg, Europeans 31.4 g/kg at the end) (Adam et al., 1953); and this difference was not due to differences in dehydration. Throughout the acclimatization period there was little difference in the rectal temperature and pulse rate of the two groups. It seems that the Asians succeeded in adjusting to the stress of a hot environment with a marked economy of water turnover. An important difference between the two groups lay in the amount of chloride secreted in the sweat; the Asians consistently lost a much lower amount (1.4 mg/kg weight at the beginning, 2.5 mg/kg at the end of acclimatization) than the Europeans (8.9 and 9.3 mg/kg respectively). This may possibly be due to low salt consumption, but it may also indicate an adaptive response, since the amount of aldosterone (which decreases chloride excretion) excreted in the urine increases in the heat (MacPherson, 1960).

In Nigeria (Ladell, 1951) Africans at first examination had an inherent degree of acclimatization, as measured by the relationship of sweat rate to rectal temperature, equivalent to that acquired by Europeans after three days of artificial acclimatization. When fully acclimatized, the Africans started sweating at lower rectal temperatures and they produced more sweat per unit surface area than Europeans (by contrast with the Asian response). In long continued stress conditions, however, the African sweat rates declined more rapidly, presumably through sweat gland fatigue, but Africans were less liable to heat collapse and to heat cramps in severe stress. Again, the chloride content of African sweat was low. Thomson (1954) pointed out that under moderate stress conditions continuing for one hour, West African sweat rates (.18 liters/m^2/hour) were lower than in European soldiers tested at the same time (.25 liters/m^2/hour) and their final rectal temperatures were also lower; and he argued that the Africans were therefore at a thermoregulatory advantage, unfortunately without considering the possibility of differences in acclimatization.

British soldiers in Nigeria were apparently more fully acclimatized than local Africans, probably because of their greater physical activity; the extent to which the African group was representative of tribal dwellers, however, is not clear. Bushmen under moderate stress appear to be as well adapted as fully acclimatized Europeans.

In the Rand mines, experienced Bantu miners had slightly lower sweat rates and slightly higher pulse rates and rectal temperatures than acclimatized Europeans in London—interpreted as a somewhat lower state of acclimatization among the Bantu, since the sweat rate could be raised to European levels by further acclimatization. In severe conditions of high humidity and low air movement the Bantu sweat rates were below those in London subjects, possibly because of sweat gland fatigue. Another study of Bantu mine workers showed that fully acclimatized Bantu had similar sweat rates to British at moderate stress, but under more severe stress conditions the African rates of sweating were lower; again this may be due to sweat gland fatigue (Weiner, 1950; Wyndham et al., 1952). The American Negro showed less deviation from the normal core temperature when working under hot-wet conditions than a white control sample matched for height, weight, and fat.

Wyndham (1970) reported rectal temperatures, heart rates, and sweat rates in a series of experiments in whites (in South Africa, North Australia, and the Sahara), South African Bantu, Bushmen, Australian Aborigines, and Saharan Arabs. The Australian Aborigines showed similar remarkable economy of water turnover to that reported in the Singapore Asians, with sweat rate levels similar to those of unacclimatized Bantu. The Saharan Arabs were intermediate between unacclimatized and acclimatized whites, and were similar to active French servicemen in the same localities and active white Australians. The Bushmen were variable, resembling unacclimatized Bantu in rectal temperature but were intermediate in sweat rate. It emerges, not unexpectedly, that the levels of physical activity play a significant role in reaction to heat stress. For heat adaptation Wyndham regards as particularly important the climates in which the groups live, and indeed concludes that life in a hot climate confers about 50 percent acclimatization by comparison with that achieved artificially.

In confirmation of this view, Hanna (1970a) examined in a hot chamber the subjects who had participated in his outdoor desert experiment, and found that natural acclimatization to desert heat does not confer full acclimatization to hot humid conditions. But other factors also enter, for whereas paired comparison of Yaqui and Mestizo showed no difference between them under desert conditions, in the hot chamber 4 out of 10 Mestizo, but only one out of 7 Yaqui, failed to complete the experiment. Similarly, a higher state of natural acclimatization does not fully account for the differences in performance between Shibipo Indians and Mestizo in the Amazon forest (Hanna and Baker, 1974).

With regard to heat stress, then, the evidence, though to some extent conflicting, appears somewhat in favor of an advantage to tropical peoples in terms of water and chloride economy and lack of ill effects, though it is not yet possible to assess what part, if any, morphological differences play. Much depends on whether the lower sweat outputs reported are due to water economy or sweat gland fatigue; the lack of cramps and collapse appears to favor the former. If sweat gland fatigue is as important as criticism of these studies implies, it might be expected that Africans would possess in compensation a greater number of sweat glands; yet this is not so. There is no evidence of any racial difference in count of sweat glands or in their pattern of body distribution (Collins and Weiner, 1965; Roberts et al., 1970). There may, however, be some qualitative difference between Negro and European sweat glands, while long-term acclimatization to heat appears to increase the maximum number of glands that can be induced to function in heat stress (Knip, 1975).

Cold

Experimental investigations of cold adaptation in which an ethnic variable has been introduced seem to have developed mainly since the report that black Americans in Korea were more susceptible to cold injury than were Caucasians, and subsequent observations elsewhere (Schumann, 1952); in an army exercise in Alaska in 1954, 50 percent of the cold injuries were in blacks, who comprised only 10 percent of the troops. The cold-exposure experiments, in which representatives of various ethnic groups were induced to submit to metabolic and temperature examinations while sleeping in cold conditions, indicate different reactions among populations, though it is difficult to quantify the effects of sleeping hard as distinct from sleeping cold on the European controls, and slight variations in the circumstances of the field inquiries are difficult to discount to some extent. Whites show metabolic compensation by shivering and grosser body movements, and so maintain warm core and peripheral temperatures; Australian aborigines allow their extremities to cool without raising the metabolism (Scholander et al., 1958). Like the Australian aborigines, the nomadic Lapps and the Arctic Indians also showed a reduction in core temperature, though here there was no difference from the controls in peripheral temperature (Irving et al., 1960). The Alacaluf Indians of southern Chile showed a slight drop in rectal temperatures to a plateau, a drop in metabolism interrupted by intermittent shivering, and a drop in skin temperatures; and though no control data were obtained, the plateaus were respectively similar to, higher than, and lower than those expected for Europeans (Hammel, 1964; Elsner, 1960). It seems that the essential difference from Europeans lies in the degree of cooling these populations can tolerate before there is any increase in the metabolic rate. Acclimatization complicates this simplification; at the end of six weeks Norwegian students sleeping cold maintained higher skin temperatures on the extremities when exposed to cold than they did initially.

In exposure to cool conditions, rectal and skin temperatures of Eskimos remained slightly above those for both black and white Americans, and their metabolic rate remained higher. In all three groups metabolic rate began to rise at the onset of shivering, at skin temperatures of 29.5°C in Eskimos and whites, and 28°C for black's; i.e., the blacks showed a delay before their metabolic rate began to rise. When groups of American blacks and whites similar in average height and weight were exposed to cool conditions (50°F, 50 percent relative humidity), after 60 minutes the black mean body temperature began to fall below that of whites. Such a difference, for both skin and rectal temperatures, was not due to metabolic differences; it could be attributed partly to differences in the thickness of subcutaneous fat, but the greater part of the variation remained unexplained, and it was suggested that different distributions of subcutaneous fat may also be partly responsible (Newman, 1969). Such would also account for some of the differences between male and female Peruvian Indians in response to cold (Hanna, 1970b), where for example, females were able to maintain higher rectal temperatures than males for a given length of exposure. A similar correlation of skin temperature with skinfold thickness seems to hold in further work on Australian Aborigines by Hammel (1964), and others, for those subjects for which individual data are given, and is suggested in the Alacaluf data.

In an attempt to distinguish ethnic from acclimatization factors in cold response, Chinchero (Andean) natives were compared with Cuzco students of European and Quechua ancestry (Baker, 1966). Their response to total body cooling (nude exposure to 14°C for two hours) was examined. The hand, foot, and mean weighted skin temperatures in the natives and their rectal temperatures were higher than in the two student groups. The Indian students had clearly lost some of their acclimatization to cold. A second investigation tested peripheral responses to extreme cooling by immersion of the fingers in water at 0°C for one hour. Only the rewarming cycle showed statistically significant differences between the groups, the Chinchero showing the shortest time before the first rewarming cycle. The finger temperatures tended to be warmer in the Indian groups than in the white, and the differences between the Indian groups despite their cultural difference appeared to be less than between Indians and whites. In another study of total body cooling, high-altitude Peruvian Indians experienced a greater initial heat loss than low-altitude Indians or whites, almost entirely as a result of high temperatures in the extremities due to higher blood flow, and these population differences were established by at least the age of ten years (Baker, 1969).

The influence of morphology on finger-cooling in unadapted subjects was examined by Steegmann (1974). Finger-cooling by wind chill in Japanese and Europeans born and raised in the tropics showed similar distributions of finger temperature in the two samples, but 4 out of 24 Europeans were withdrawn from the test on account of incipient cold injury and 1 out of 33 Japanese was withdrawn. The finger temperatures for each individual over the exposure periods 6-30 minutes and 40-70 minutes were examined in relation to various body measurements. In both the Japanese and Europeans, the finger temperature during exposure was positively correlated with that at the start of the experiment. In the Japanese, the finger temperature was positively correlated with skinfold thickness measured on the chest and abdomen, and in both Japanese and Europeans, with relative sitting height. The former seems to reflect the higher core temperatures that the better insulation helps to confer, and the association with sitting height is also compatible with this explanation.

These studies suggest that there are indeed ethnic differences in the ability to withstand cold stress, and that these are partly, but probably not completely, attributable to differences in morphology and body composition and in acclimatization. Certainly it seems that short-term acclimatization to cold in Negroes operates in the same direction as in Caucasians, though there remain differences after this is taken into account. While it has been shown that tropical subjects who have never known seasonal cold respond differently to cold exposure, this may be due to uncontrolled features (e.g., diet) of the experiment. Too few variables, however, have been considered for any final conclusions to be drawn. For instance, little attention has been given to the possibility that differences in physical fitness are responsible for the apparent differences in cold adaptation, for it has been shown that physical fitness training increases skin and extremity temperatures and diminishes core temperatures when the subject is exposed to cool conditions.

INDIRECT EFFECTS

Indirect factors, as well as the more direct relationships already discussed, also enter into the somatological relationships with climate. Take, for instance, the effect of cold on manual performance. Accidents increase as temperature moves toward the extreme. At low ambient temperatures, particularly if there is also wind-chill or if there is local cooling of the hands, there is probably some increase in reaction time of hand movement; tracking tasks are less well performed; general manual dexterity diminishes particularly with local cooling of the arm and not necessarily with body cooling; tactile discrimination diminishes markedly as finger temperatures fall below about 40°F; and limb strength is impaired during cold exposure by an amount depending on the drop of local temperature of the muscles below about 80°F.

A number of tests of manipulative ability were made on Eskimos and Arctic Indians by comparison with a group of white males resident in the Arctic for some six weeks. In every measure the native groups were superior to the Europeans; at comparable temperatures the decline in each function for the natives was only about half that for the whites. This may be due to physiological acclimatization, through the medium of increased blood flow through the hand and fingers (Egan, 1963); resting hand blood flows and skin temperatures of Eskimos are considerably higher than those of Canadian students, and the same measurements in Lapps are higher than in control Norwegian students, while Lapps also have earlier onset of cold vasodilation. In the Quechua, sex differences in hand cooling brought about by the women's daily tasks involving the immersion of the hands in cold water (preparation of meals and washing of clothing) may induce differential acclimatization. Similarly, in Korea, the women divers have greater overall body insulation for a given subcutaneous fat thickness than nondivers in the same community (Paik et al., 1972). This appears to be due to vascular adaptation, for no hunting reaction was observed at 6°C; and the findings suggest a greater degree of vasoconstriction of the peripheral blood vessels, giving a lower heat loss from the limb and a greater fraction of the hand blood flow return via the superficial veins.

A similar mechanism may be responsible for the report that the Arctic Indians have superior ability to keep their hands warm in cold water. That short-period acclimatization is not the only factor is shown by the difference in response after training between north and south Japanese

(Yoshimura and Iida, 1952); perhaps a developmental factor is involved. A genetic component is suggested by the striking similarity in response that occurs in twins (Meehan, 1955), and by the differences between U.S. white and black soldiers reared in similar climatic environments (Iampietro et al., 1959). When comparisons were made of American blacks and whites during 45-minute finger immersion in a water bath at 0°C, the average black fingertip temperature was significantly below that of the whites (2.7°C as compared to 7.2°C), while rewarming began at a much later stage among the blacks than among the whites (15.9 minutes and 9.2 minutes respectively) and some blacks showed no rewarming at all. Alacaluf Indians experienced no pain or discomfort, in marked contrast to whites, when the feet were placed in cold water after a period in warm water.

There are also indirect effects at the opposite extreme, of heat. Physical work can be done by fit and trained individuals in extraordinarily severe heat conditions—for a period. Indirect effects begin to occur at much lower levels of stress. There have been many investigations of performance of skilled tasks at different temperatures. Deterioration in performance begins and mistakes begin to increase rapidly once the effective temperature rises above about 83°F, though skill appears to delay the onset of deterioration in accuracy. Moreover, motor nerve conduction velocities may be involved. Among East Africans, in subjects with higher skin temperatures the conduction velocity was greater by 2.6 m/sec/°C (Thies et al, 1967). Another indirect factor is age; older men working in heat show greater strain, manifest in their higher rectal temperatures and the faster rate at which these increase with time than in younger men (Lind et al., 1970; Wagner et al., 1972), perhaps through some age limitation in the sensitivity and secretory capacity of the sweating mechanism. Any somatic difference that would improve reaction time or dexterity in an extreme climate is likely to have been of selective significance.

PART II

Developmental Aspects of the Association

The review of the distribution of variations in adult human physique suggests that these may be selectively advantageous in different climates. This suggestion is supported by the experimental evidence, which points to the existence of morphological and ethnic effects, both direct and indirect, on climatic stress tolerance. The situation is far from simple; there are complications of acclimatization, and of marked differences in the reactions of individuals, particularly at high levels of stress. Nevertheless, the indications are strong.

The problem therefore arises as to how this association of physique and climate comes about. Does it represent genetic adaptation in its entirety? Is it purely phenotypic, a total interference with a common pattern of development by the environment in which the individual grows? Or does it represent an interaction between these two sets of factors? There are several lines of evidence.

PHYSIQUE IN CHILDHOOD

The association of adult physique with climate is foreshadowed in preadult life. While there has so far been little investigation of the physique/climate relationship during development, the few studies of body dimensions in children that exist point to the same tendency to linearity of form in warm-climate children as occurs in adults. The majority of relevant studies concern physique assessed in terms of weight and height. Certainly, the relationship of mean weight with mean temperature holds throughout the later growth period. The association is complicated by age variations in weight and height, but these difficulties may be met by fitting smoothed curves for log weight and stature by age. This was done for 150 samples of indigenous children in different parts of the world, and from these the means at ages 6, 10, and 14 were established for each sex (Roberts, 1960). The correlation of log weight with mean annual temperature of the habitat was calculated at each of these ages, the zero-order correlation coefficients being, for both sexes combined, -.706, -.587, and -,525 at ages 6, 10, and 14 respectively. The partial correlation coefficients at similar levels indicate that the association of weight with temperature is not attributable to stature differences, and the multiple correlation coefficients show that up to 78 percent of the total variance in mean weight at each age is attributable to variation in climate and stature in combination. Again, no causal relationship is claimed;

this is merely an empirical association which is of interest whether it is direct (i.e., physiological) or indirect (due to nutritional differences or genetic adaptation).

An association of weight and climate may be traced further back in development. Many records of birth weight in indigenous groups are found in the literature, and these show much more variation than one would perhaps expect; the highest mean birth weight in the sources listed by Meredith (1970) is approximately 160 percent that of the smallest. A number of environmental factors affecting birth weight have been identified; climate is among them (Roberts, 1969).

When a similar empirical correlation analysis is applied (Roberts, 1969), the linear correlation coefficient of mean birth weight in relation to mean temperature is -.76, though a curvilinear association would be equally acceptable. Differences among the means of the major continental groups of man remain highly significant after the effect of temperature has been taken into account; and again within the continental groups the correlation with temperature remains highly significant (Fig. 7). Obviously, in this case there is not likely to be the same functional reason for the climatic relationship as in adults or older children, for the fetus is protected from external temperature and other stresses by the constancy of the intrauterine environment. Yet the fetus does produce heat, and the excess is dissipated via the mother, which she in her turn must lose. In situations where heat loss is difficult, it may well be advantageous to carry a fetus in which the amount of heat-producing tissue is reduced. This again may well be brought about by physiological mechanisms, a direct growth rate response by the mother to

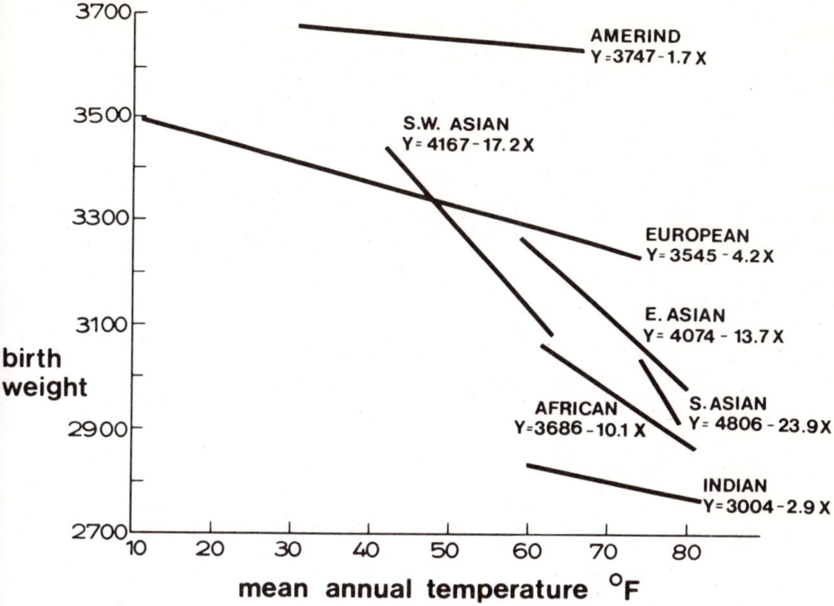

FIG. 7
Intravarietal regression of mean birth weight on mean annual temperature.

external temperatures which would be compatible with the evidence for slight seasonal variation in birth weight.

Examination of birth weight in Hong Kong in relation to meteorological variables of prenatal months and the birth month, identifies an appreciable effect of temperature of birth month, an effect greater than that of maternal age (Roberts, 1975). It seems that babies born in the hotter months are smaller and hence, since there is no association of gestation length with any of the meteorological variables, that the intrauterine growth rate toward the end of gestation may be slower in the hotter months. This would point to a physiological mechanism—a direct birth weight response to external temperatures via the mother.

There may be a further advantage in hot conditions in that a lighter fetus would tend to diminish the total weight that a pregnant woman has to carry and hence the amount of work in any movement. However, the climatic association may be spurious, the lower birth weight of tropical peoples being due to their generally lower standard of living, with which unsatisfactory hygiene or diet is so closely associated. All that can be said is that the possibility of some genetic contribution cannot be discounted.

The extent to which the weight variations (associated with temperature) in later life derive from the birth weight variations, or to which the birth weight variations can be said to anticipate the adult weight variations, is by no means clear. Within a normal population there is little correlation of weight at birth with weight in later life. If there is an adaptive genetic component to the adult weight/temperature association, and if there is a similar component to the birth weight/temperature association, then a dual system of selection must have operated, for those genes that affect prenatal development do not necessarily affect postnatal growth or adult size. But again there is complexity, for genetic control of variation within a population implies nothing about the extent of genetic control of variation between populations.

PHYSIQUE AND NUTRITION

While nutrition may well contribute to some degree to the padded physique of Arctic peoples, there is some evidence that extreme tropical physiques owe little to immediate dietary differences. This comes from an alternative method of physique assessment, somatotyping, a procedure which summarizes rapidly, in terms of three components (endomorphy, mesomorphy, and ectomorphy), an impression of physique gained from many measurements and their interrelationships. There are relatively few samples available of adult indigenous peoples, but they range in habitat from the Nilotes, who are subject to a climate extreme in heat and seasonal humidity, to the Eskimo of Alaska.

In endomorphy the modal male values diminish from 3.9 in Eskimo, 3 in Japanese, 3 in English, 3 in Indian, 2 in East African, 1.6 in Manus, to 1 in Nilotes. This gradient probably represents a diminution of the amount of fat in the body, since there occurs a high correlation (+.853) in whites between the amount of body fat and the endomorphy rating. This interpretation is supported by the fact that, by contrast with cold-climate peoples, subcutaneous fat is low in tropical peoples and minimal in Nilotes, while studies of body composition in other tropical groups show a low percentage of total body fat. As intimated previously, a low fat content would be advantageous in a hot habitat; the diminished insulation of a minimal subcutaneous layer would facilitate heat loss, while less fat in the deeper tissues would reduce the weight, the work performed, and heat produced in physical activity involving body movement. It is possible that the body fluid balance is also to some extent involved in the endomorphy gradient. The variations in mesomorphy cannot, it seems, be linked to any climatic factor. The gradient in ectomorphy is again striking, with modal values of 1 in Japanese, 1.8 in Eskimo, 3 in East Africans, 3.7 in English students, 5 in Nagpur, and 6 in

Nilotes, though it is interrupted by a low value in Manus (1.6) presumably due to the very high mesomorphy (6.7). Since there is no significant relationship of ectomorphy to specific gravity or body fat, extreme ectomorphy, which maximizes the body surface area relative to the body mass, is of obvious adaptive value in an environment where heat stress is not uncommon for an active individual. The variations in two components are thus compatible with the adaptive climatic relationships indicated by anthropometric data (Bainbridge and Roberts, 1966; Roberts and Bainbridge, 1963).

It is well established, however, that human physique shows physiological plasticity, i.e., phenotype modification, particularly with health and nutritional variation. At first sight the differences between the Nilotic and European component modal values are in the very direction observed in the Minnesota starvation experiments, where with fairly severe semistarvation involving average weight loss of 24 percent, endomorphy and mesomorphy diminished on the average 1.7 and 1 point respectively, and ectomorphy increased 2 points. This may seem to suggest undernutrition rather than adaptation as the cause of the Nilotic attenuation of physique. But if this were so, the Nilotes would also need to be more poorly nourished than the tribal Kikuyu living under crowded reserve conditions, while the magnitude of the modal differences from European values would imply a severe degree of semistarvation, which clearly did not occur in the Nilotes. Further evidence comes from comparison of the Nilotic tribal subsamples, of Shilluk and two Dinka groups (Ruweng and Ageir). A genetic survey showed that these groups were homogeneous in allelic frequency at a relatively large number of loci. Hence their physical differences are to be attributed primarily to the effect of environmental differences on the phenotype. Such environmental differences exclude climate, since it varies very little within their habitat, and relate primarily to their nutrition or mode of life. Despite the differences in mode of life, the somatotype differences between the subsamples are relatively slight but are in the direction expected from their diets; the poor, largely carbohydrate diet of the Shilluk would tend to contribute to the formation of fat, rather than bone or active tissue, and the largely protein diet of the Ruweng would tend to diminish fatty tissue rather than to increase the other constituents. The mixed diet of the Ageir, containing protein-sparing carbohydrate as well as adequate to high protein, would tend to produce increased size in weight and probably height. These are the differences that are in fact found. Comparison of the Nilotic tribal subsamples, then, which experience a common climate but differ in nutrition and mode of life, indicates that variations in physique may be related to the availability of nutrients during growth. But the direction and minor extent of these variations show that the difference of the Nilotic physique from that of other populations is not to be attributed to these factors.

CLIMATE AND GROWTH PROCESSES

The possibility that the physique relationship represents an immediate effect of climate on growth processes requires consideration. There is so far no direct evidence of such an effect in man, though it has been reported in animal studies and there are hints from a number of sources.

First there is the limited evidence from groups growing up in environments to which their forebears were not indigenous; some have been resident for several generations, others are newly migrant. White Americans in the United States show a correlation of weight with climate, which cannot be attributed to long genetic selection. However, the regression coefficient of weight and mean annual temperature is only about one-third that observed in the indigenous series, which suggests that only a portion of the indigenous weight/climate association is phenotypic; Newman and Munro (1955) were inclined to attribute the association to an increase of appetite in cold climates. Among American college students there is a clear tendency for those in the warmer south to show greater lower limb length. As for new migrants, European children raised in interior Queensland tend to be lighter for their height, while Japanese raised in tropical Hawaii show an increase in ectomorphy but no change in endomorphy or mesomorphy. All studies of migrants from one environment to another in comparison with their kin remaining at home demonstrate a tendency for stature and other longitudinal body dimensions to be greater in the new environment. This finding has come from studies of Europeans in New York, Japanese in Hawaii, Chinese and Mexicans in the United States, and Japanese and Italian Swiss on the Pacific coast (Hulse, 1957; Lasker, 1946, 1952, 1954; Goldstein, 1953; Greulich, 1957). These studies clearly suggest the occurrence if not the extent of human plasticity, perhaps through some direct influence on growth. Other possible explanations are that migrant subgroups may be different from their parent stock and that those with a particular growth pattern tend to choose the habitat to which they are most suited, while the meticulous studies of Hulse (1968) suggest that heterosis, as a result of change in mating patterns, may be partly responsible. The tendency to stability of body proportions, on the other hand, is well documented. A number of studies in the southern United States show clear differences between blacks and whites living under similar climatic conditions, with hybrids showing intermediate values. Shapiro's 1939 analysis of Japanese migrants in Hawaii demonstrated "that although the Hawaiian born are taller and have larger sitting heights than the immigrants, nevertheless the proportion remains unaltered." Thus populations may tend to retain their relative sitting heights in different environments, and different populations living side by side under identical climatic conditions may all retain their characteristic values. The third-generation Japanese migrants in the United States retain not only their characteristic relative sitting height and surface area, but also remain intermediate between American and Japanese in chest measurements and pulmonary parameters such as vital capacity, aerobic capacity, and ventilatory volume (Raven et al., 1974). It seems that the relationship of relative sitting height and temperature is not due simply to modification of the phenotype, though this may contribute to it.

Skeletally, a direct climatic influence on growth would imply that tropical residence might delay epiphysial fusion of the long bones so that longitudinal growth would continue over a longer period. There is indeed a suggestion from rather crude extrapolation of height and weight curves to adult levels in Africans and Europeans that this does occur, with the extreme Nilotes being perhaps 1½ to 2 years behind the European age norms in attaining their adult size (Roberts, 1961). Radiographic studies of indigenous children in the tropics indicate that later epiphysial fusion of the carpal region indeed occurs, but it is not possible to extrapolate from carpal ossification to long-bone epiphysial fusion, for the correlation in ossification time is far from perfect. In any case both these aberrations may represent intervarietal rather than climatic differences. In whites, though later maturing individuals are higher in ectomorphy, only a small fraction of adult linearity of form is due to the more protracted growth of the long bones before the epiphyses fuse; linear people just develop late and are less advanced in their growth at all ages (Tanner, 1962). It may be that the contribution of longer continued growth to the Nilotic somatotype is somewhat greater than in Europeans; but since linearity is clearly detectable in Nilotic children, it seems more reasonable to attribute only a small portion of the Nilotic somatotype to prolongation of growth, whether or not this is due to climatic influence.

Another way in which climate may have a direct effect would be be accelerating longitudinal growth per unit time. For this there is little evidence either way. Slight seasonal variation in growth patterns is well established in temperate regions, with growth in stature predominating in the warmer months; this may be related to general health, to different seasonal metabolic requirements, or to other variables. Tropical residence may or may not cause European children to be a little more spindly in physique, but the fact that

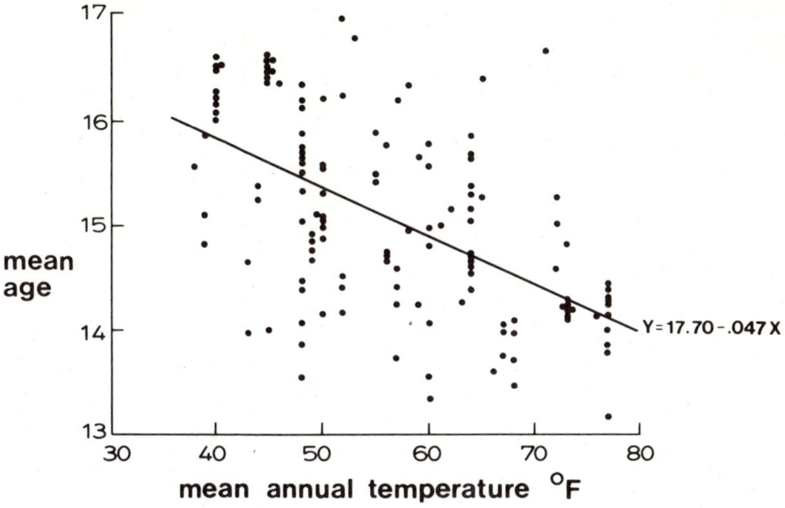

FIG. 8
Menarcheal age and temperature (Pearl's series).

they do not develop a Nilotic physique suggests again that this owes little to any direct climatic influence on growth rate.

There is one feature of maturation which shows a climatic effect. When the mean menarcheal age in girls is plotted against the mean temperature of the area of their habitat, there is a clear indication of a climatic association, the correlation coefficient r = -.567 and the regression coefficient b = -.047 being highly significant (Fig. 8). There are many sources of bias in using such recall data, but confirmatory evidence comes from status quo surveys analyzed by probits (Roberts, 1969). Within continental groups the correlation of menarcheal age with mean temperature is appreciable (r = -.40), and there are significant differences between continental groups. The climatic effect within them may well be physiological, perhaps related to the environmental association of basal metabolism, some of the oxygen required for growth and development being diverted to maintenance of body temperature in cold climates. The direction of this effect argues against the suggestion of climatic delay of termination of long-bone growth in warm climates, and for an explanation in terms of an intervarietal difference; African menarche in a given climate is later than European, and African attainment of adult stature is later. This is, however, merely a hint that the two lines of evidence are in accord.

There is much research to be done to clarify how the climatic association comes about. Much of the difficulty is due to inadequate genetic knowledge. We are ignorant of too much of the genetic basis of intrapopulation and of almost all interpopulation variability. But it seems that the physique/climate association is detectable during the growth period, and therefore is not primarily due to delayed adulthood. It is not primarily due to nutritional differences; there is some direct physiological implication in maturation. Physiological phenotypic variation and genetic differences in combination probably account for the association. Of these two, the latter appears on present evidence to be predominant, and this implies that natural selection acting on the gene pools of populations is ultimately responsible for the physique/climate association.

PART III

Other Climatic Associations

The discussion so far has centered on the association between physique and climate, and how it comes about. This emphasis merely reflects the fact that most investigation so far has been concerned with or is relevant to this relationship. There are, however, several other factors to which serious attention has been given and others about which there has been only speculation.

NOSE SHAPE

The first morphological character to be examined for a relationship with climate was the nasal index, which expresses the breadth of the nose relative to its height. There is considerable variation from people to people, and the genetics of the character remains unknown, but the variation is certainly not purely phenotypic in origin. It was Kallius at the end of the last century who first suggested that the nasal structures were probably of use in raising the temperature of inspired air, and simple experiments soon confirmed this, for example, those measuring air temperature as it passed in at one nostril at 10-12°C and passed out at the other at 31°C and saturated with moisture. Thomson's preliminary paper (1913) showed correlations of variations in the nasal index with isotherms, and this was followed by the classic paper of Thomson and Buxton (1923). Examining mean nasal index in indigenous peoples in relation to the climates inhabited, they found a correlation of the nasal index with mean annual temperature of +.629, with relative humidity of +.419, and with temperature and relative humidity together of +.724. Thus temperature and humidity were both important, and they suggested that the modifications of nose shape might be related to the need for moistening inspired air or might be related to heat loss from the respiratory tract, arguing that "in the tropics the loss of lung water (evaporative) from the respiratory tract is undoubtedly of considerable physiological importance. In air containing the highest relative humidity the existence of a free passage of entry is of advantage, in order that sufficient air may be breathed in to absorb water from the respiratory tract." Their correlation coefficients were subsequently confirmed by Davies (1932) on longer series of data.

Further evidence comes from physiological considerations. A prime function of the nasal epithelium appears to be moistening of the air. The cilia are more susceptible to changes in the consistency of the film in which they work, and therefore to drying out, than they are to quite large degrees of heating and cooling. It is, of course, the cilia which move toward the exterior the mucus and dust entering the nose; they thus clean the inspired air, which is broken up into shallower streams passing through the cleft between the convoluted surfaces of the mucous-membrane-covered conchae, and Ewert (1965) has demonstrated the strong influence of relative humidity on mucus flow rate. Consequently, a closer association may be expected between the nasal index and that physical factor primarily concerned in the exchange of water from nasal epithelium to inspired air, or from the respiratory tract surface to inspired air. This factor is the absolute humidity. Weiner (1954) therefore reanalyzed the data of Thomson and Buxton (1923), using more meaningful measures of humidity, and found a correlation coefficient with wet bulb temperature of .77 and with vapor pressure of .82, a significant increase over the correlations previously observed.

Clinical work also shows the importance of humidity. In patients with tracheotomy the nasal passages are bypassed, the air reaching the lungs is little different in temperature and humidity from room air, and it takes up moisture in the bronchi from the mucous membranes so that secretions become dry. This is not a serious handicap to anyone who can cough up viscid secretions, but it may be disastrous in patients whose power to cough is impaired and whose bronchi thus become blocked. This has become very obvious in patients with various forms of paralysis necessitating tracheotomy and artificial respiration. If in these cases inspired air is properly humidified, there are no sinister complications of encrusted secretions such as were characteristic of tracheotomized patients in the Danish poliomyelitis epidemic of 1952, before the need for humidification was recognized.

But the warming function remains important. Webb (1951) measured inspired and expired air temperatures at the front, middle, and back of the nose at different ambient temperatures. In each, the inspired air was rapidly warmed in its course through the nose. At the back of the nose it was still several degrees below 37°C, and at the same location the expired air was warmer but again below 37°C and tended to cool as it progressed outside. This temperature drop in the expired air was successively greater as the ambient air temperature became lower. Thus with nasal breathing cold damage to pulmonary tissue is unlikely under conditions as extreme as those used. But these findings are also of interest in that they demonstrate some heat conservation mechanisms—the cooling of the expired air as it travels toward the outside—so that less heat is lost through this channel than if the expired air were at body temperature. This phenomenon is more marked at lower ambient air temperatures, and represents a considerable and useful conservation of heat, particularly in very cold environments.

FACE AND HEAD FORM

Face and head form has attracted much discussion but relatively little investigation, and this only recently. In conditions where heat loss is appreciable, there is no consistent change in the superficial circulation of the head comparable, say, with that occurring at the extremities; consequently, heat loss from the unprotected head may be excessive. There is certainly a linear correlation of heat loss from the head over ambient temperatures varying between -20°C and +28°C, so that the insulation of the head tissues remains unchanged with the external temperature. Second, the head insulation remains unchanged even if there is general vasoconstriction over the remainder of the body. Thus there is no vasoconstriction in the head in response to cold, either when cold stimuli come only from the head or when the rest of the body is in a general state of vasoconstriction. There is therefore a steady flow of heat to the peripheral tissues of the head in cold conditions; its obvious function is to keep the brain temperature normal. The importance of this function is indicated by the fact that the head's response to cold is in direct contrast with what happens over the remainder of the body in cold conditions. These findings point to the possible importance of variations in the cephalic index and the retention of head hair.

Confirmatory evidence comes from investigation of the decrease in skin temperature at various sites on the head and face (Froese and Burton, 1957; Edwards and Burton, 1960). On going from a temperature of 24°C to 0°C, the greatest drop in skin temperature occurs at the tip of the nose and on the tip of the ear, followed by the ear lobe and the cheek below the eye, the bridge of the nose, the side of the face, and the chin. At 0°C the lowest skin temperatures are on the nose, the ear, and the chin, and these are the parts that are most susceptible to frostbite. The warmest parts are on the scalp under the hair, on the neck, and around the eye, and then the forehead and side of the face; these are the parts where heat loss is greatest, and hence where insulation would exert the greatest effect. The distribution of the warmer areas is closely related to the distribution of the superficial blood flow in the head. Here, then, is a mechanism by which variations in head and face form, hair form, and hair intensity may be understood.

The results provide some confirmation of the speculation of Coon, Garn, and Birdsell (1950) regarding the mongoloid face. They pointed out that in prehistoric times the cold encountered by peoples living in northeast Asia must have been an extremely selective environment. Whereas prehistoric man was probably expert in using skins for clothing, his head and particularly his face would have been unprotected, so that here occurred one of the most stringent opportunities for environmental selection of anatomical characters that has occurred in human history. This selection, they said, gave rise to the mongoloid face with its characteristic flatness, fat padding, sparse beard, and coarse black hair on the head. Their hypothesis was that selection in response to the cold conditions caused the following: The bony orbit was raised vertically to make room for a heavy fat padding round the eyeballs; the brow ridges were reduced because the sinuses therein were particularly vulnerable; the malar bones were extended forward and enlarged, so that the nasal skeleton was flush with the cornea; the nasal passages remained deep to heat the air taken in, and the width of the face allowed the nasal resonance chamber needed for speech to remain large. Three principles are involved: (a) reducing the surface area to a minimum by flattening all protuberances as much as possible, (b) padding the surface with fat for insulation, and (c) protecting the nasal passages for warming the inspired air. Again, however, the situation may not be this simple. Recent studies of the direction of flow over the body surface of air warmed by it show the importance of minor topographic variations, and the extension of this work to the examination of population differences in morphology is awaited with interest.

In one notable attempt in population comparison, Steegmann (1972) examined continuous surface temperature records at five sites on the face, during the laboratory exposure of the head to cold in 33 Japanese and 25 Europeans in Hawaii. The results were compared with a large number of head and face measurements. Not unexpectedly, cold-induced cycles of vasodilatation and constriction occurred at all sites, but especially on the forehead and nose, but the correlations of morphology and skin temperature, though low, are suggestive. In both samples, cheek temperature was negatively correlated with the thickness of the soft tissues of the cheek (Japanese -.45, Europeans -.58). The malar temperature remained higher in those in whom the malar was less protrusive (Japanese +.37, European +.36). Nose temperature increased with head size, and forehead temperature with head length. Thus, if the optimal cold adaptation is the retention of a warm surface, this would be achieved by thin cheek tissues, thin adipose covering on the face below the eyes, nonprotrusive and small malars, a relatively large or long head, and an active hunting response to selected sites. This is only partly concordant with the speculations of Coon, Garn, and Birdsell. But of course it is by no means clear that continued high surface temperature is the optimal form of cold adaptation. So long as there is an adequate hunting response to prevent actual tissue destruction, efficient insulation seems an important requirement.

Of the internal structures affecting face and head form, sinus size may be involved in climatic adaptation. The function of sinuses is still unclear, and several suggestions of climatic relevance had been made, e.g., the maintenance of internal cranial temperature (Maurer, 1953), the humidification of inhaled air (Eckert-Mobius, 1933; Negus, 1957); other suggestions include resonance and decrease of skull weight (Allen, 1961). Koertvelyessy (1972) noted that the surface area of the frontal sinus in Eskimo from different localities was significantly less in those from the colder localities.

SKIN PIGMENTATION

Of all features which vary visibly from person to person and population to population, skin color is the most striking. On this topic there is more discussion and speculation than satisfactory investigation, primarily because it is a character difficult to measure. Much early discussion is based on verbal description of matching against standards, and sophisticated analysis had to await the quantification of color by reflectometry.

Skin color is due to the presence of the pigments melanin, melanoid, hemoglobin and oxyhemoglobin, and carotene, and to an additional optical effect, "scattering" (Edwards and Duntley, 1939). Melanin, the most important pigment, is located as discrete particles mainly in the deepest layers of the epidermis at the base of the stratum of Malpighi in light-skinned peoples; in dark skin the particles are more abundant, spreading to the greater part of that stratum and being more diffused in the superficial layers. The individual granules vary in color from yellow to black, and it is their number that determines skin color. The normal deposition of melanin is genetically controlled, and there is also short-period variation, in response to exposure particularly to ultraviolet light and also in some physiological and disease states. Melanin is synthesized by the melanocytes from the amino acid tyrosine by means of tyrosinase, and the process is under the control of the melanocyte-stimulating hormone. The total number of melanocytes per unit area of skin is the same in all races on which quantitative histological studies have been carried out, so that dark skins owe their pigmentation to increased melanin synthesis. Melanoid, derived from disintegrated melanin particles, imparts a rather yellowish component to the skin, and its amount depends on the melanin concentration and the thickness of the epidermis in which the pigment is stored. Hemoglobin, the oxygen-carrying pigment in the blood, shows its effects least in dark skin, being obscured by the melanin in the surface tissues overlying the blood vessels. Carotene is a coppery light brown, occurring in the dermal and subcutaneous fat and in the thicker areas of the stratum corneum; it is derived from dietary sources.

The general association of dark skin and tropical habitat, especially in regions of more intense solar radiation, has long been known, though the geographical distribution in terms of reflectance values is only now emerging; over 80 percent of the total interpopulation variance in reflectance at each wavelength is attributable to climatic variables (Roberts and Kahlon, 1976). The distribution, which seems to accord with the general mammalian pattern summarized in Gloger's rule,[2] suggests fairly strong selective advantages in variations in skin color. That parallel gradients are observed in different continental areas suggests that darkness of the skin may have evolved several times in man, again pointing to the importance of dark skin in low latitudes. The geographical distribution suggests an association with the amount and intensity of insolation. The ultraviolet parts of the spectrum in particular, while at minimal exposures they have some beneficial effects, at greater doses promote injury in the form of sunburn and carcinogenesis. The carcinogenic wavelengths lie between 2537 and 3341 Å, while erythema is induced increasingly rapidly at wavelengths of less than 3200, reaching a maximum at 2800 Å. However, the relationship of pigmentation to solar radiation is not simple (Blum, 1945).

First, man is protected by the earth's atmosphere from much ultraviolet radiation; indeed, most radiation below 2850 Å is absorbed by the ozone. That protection is lower in the tropics, since, though details of the geographical distribution of ultraviolet remain to be established, the shortest passage through the atmosphere occurs when the sun is directly overhead. In temperate latitudes the additional thickness of atmosphere to be traversed diminishes the intensity of ultraviolet radiation severalfold. There is, in addi-

---

[2]Gloger's rule states that intensity of melanin pigmentation tends to decrease with mean temperature, that in high temperatures high humidity promotes formation of the black eumelanins and aridity the yellower phaeomelanins, but such differentiation by humidity does not occur at low temperatures.

tion, considerable radiation from sky reflection, while in industrial areas atmospheric pollution absorbs a great deal. Second, much of the radiation reaching the surface of the body does not reach the deeper melanin-containing layers. It is scattered, reflected, and absorbed in the outer horny layer of the skin (though in dark skins this layer, too, contains some melanin). However, some radiation reaches the deeper layers. When sheets of the stratum corneum from Africans and Europeans in Nigeria, shown to be similar in thickness to each other, were placed on a photographic plate and exposed to ultraviolet irradiation, the European specimen showed pronounced penetration, whereas the degree of penetration in the darker African skin was much less. Moreover, the positive results of ultraviolet therapy in whites with rickets show that active rays do penetrate deeply enough to affect at least the superficial capillary circulation or to stimulate the nerve endings. Other experimental studies of differential penetration have shown that while transmission of invisible rays is reduced, wavelengths of 3650 Å penetrate white human skin 2.2 mm thick. Pronounced penetration into the corium and subdermal layers requires rays of over 4000 Å. A considerable proportion of ultraviolet rays at 2500 Å reaches the corium, but rays at 2700-2800 Å are practically all absorbed in the epidermis, and Negro skin absorbs particularly in the region from 2537 to 2800 Å. The barrier of melanin in the basal epidermal layers is obviously effective.

There are other lines of evidence besides the experimental. The fate of those dwellers in the tropics who have little or no melanin is instructive. Adult albinos in tropical areas develop a thickening (absent in young children) of the horny layers of the skin, a less efficient barrier than an abundance of melanin. Second, there is the evidence of cancer development. The rarity of cutaneous cancers in dark-skinned as compared with light-skinned subjects has often been noted. Those Europeans who are affected tend to be outdoor workers whose face and dorsum of hand (exposed areas) are the most frequent sites, while in whites in the United States there is a latitudinal north-to-south gradient of increasing development of these tumors. The selective effect of protection against these tumors is likely to be small, for they are relatively infrequent and they occur mainly in the later years of life after reproduction has ceased. Another type of tumor is the melanoma, also rare but selectively more severe since it can occur in young adults; this is a pigmented tumor which in Europeans arises from pigmented spots in the epidermis, and in Negroes from relatively unpigmented regions of the body. While cases are rare in Europe, there were 395 melanoma cases in 20,000 white Australian patients, all of whom had lived a long time in Australia; this high prevalence could well be related to excessive exposure to sunlight. Melanomas are said to be rarer in Negroes, yet malignant melanoma is, next to squamous-cell cancer, the commonest form of malignant disease seen in east and central Africa, and in Bengal it is much more common than in Europe. Most of the squamous-cell cancers are due to malignant changes in chronic tropical ulcers of the legs and the epitheliomata associated therewith, while the malignant melanomas occur mostly on the foot on its unprotected plantar surface. The extent to which radiation is involved, in that injury exposes the deeper tissues to it, is not clear. There is, however, considerable confirmatory evidence from animal experiments on the relation of pigmentation to cancer.

It is possible that pigmentation may be related to heat exchange. Black skin absorbs 80 percent or more of incident visible light, while light skin absorbs only about 60 percent or less. Black and white skin reflectance curves differ less at the shorter infrared wavelengths (700-1000 μ), though again the black absorbs slightly more. It is therefore to be expected that black skin would be heated more than white skin by the absorption of sunlight. The expected energy absorption has been calculated for white and for black skin exposed to radiant energy of similar spectral distribution to sunlight at the surface of the atmosphere; relatively untanned black skin absorbed 34 percent more energy than white skin. Since the passage of sunlight through the atmosphere filters out more energy in the infrared region than in the visible, the contrast in heat load from the visible spectrum may well be greater by the time sunlight has reached the earth's surface. In the American desert the additional heat absorption of unclothed white men exposed to sunlight amounted to 140 calories per hour on the average, a considerable heat load when it is remembered that the resting heat production is usually taken as 90 calories per hour. Under hot desert conditions, therefore, the additional heat absorption of a Negro may be 40 calories per hour or more, a potentially heavy strain on the heat-regulating mechanism under conditions of severe stress. Under these conditions the maintenance of an efficient sweating mechanism is critical. While it may be thought that dark skin would not only absorb but also radiate heat more easily, thus facilitating cooling, experiments so far show that radiation from the skin surface at a temperature of about 35°C is in the long infrared portion of the spectrum, where white skin behaves as an almost perfect radiator; thus the color of the skin is not an important factor in promoting radiant heat.

On exposure of light-skinned people to sunlight, damage to the germinative layer of the epidermis causes the release of substances which dilate the superficial vessels; when exposure continues over a long period, edema and other unpleasant effects follow. Mild sunburn induces an increase in pigment formation and a migration of melanin from the basal cells toward the surface (suntan), and the pigment once formed persists. With this the erythema threshold is raised, and may remain raised for a period of weeks; thus there is a protective effect stemming from thickening of the stratum corneum and melanization. Acclimatization to sunlight has obviously occurred. Dark skins also experience tanning but are much less susceptible to sunburn damage than are light skins. Sunburn damage to the skin involves the sweat glands, and with severe sunburn their functioning and hence heat regulation may well be disturbed. Under conditions in which efficient sweating is critical, the survival value of any feature (such as pigmentation) which may prevent sweat gland damage will be great.

While there is little doubt that skin color variations in human populations have been produced by selection, the selective advantages of a dark skin in a tropical environment are obviously complex and still not fully clear. But also selection for fair skin in higher latitudes may well have occurred. In northwest Europe the conditions of coolness, heavy cloud cover, and relatively little radiation stress are not those that would call for dark skin on any of the hypotheses outlined above. The occurrence of the lightest skins here could perhaps be due to the effects of chance on Europe's ancestors, implying selective neutrality of lightness. However, the suggestion from recent genetic studies of overall dominance of the genes from the white parent over those from the black parent carries the implication that selection for depigmentation may have been strong. The ultraviolet rays penetrating the skin cause a rearrangement of the atoms in the sterol molecules, endowing them with the properties characteristic of vitamin D, and this antirachitic potency is most effective in the range 2500-3130 Å. In conditions of minimal sunshine where dietary vitamin D is low, increased access of these rays to the deeper skin layers could well be selectively advantageous in preventing the development of rickets. There is little direct evidence, but it is interesting that black children dwelling in the great cities in the northern part of America and Pakistani children in Glasgow are more susceptible to rickets than are white children; and it is possible that this may indicate something more than a dietary deficiency. It may be due to the fact that, of the lessened amount of ultraviolet light reaching the skin in the temperate latitudes, again diminished by atmospheric pollution and indoor living, what remains after absorption by the melanin of black skin is insufficient to stimulate the production of vitamin D in the dermal layers.

But perhaps skin color itself represents a complex of entities for which a variety of differing selective advantages and disadvantages are to be sought. For the environmental factors with which reflectance is correlated are not constant over all wavelengths, and in particular appear to distinguish the longer from the shorter (Roberts and Kahlon, 1976), suggesting that the components of skin color contributing differentially at these wavelengths may have different selective values.

Conclusion

In drawing together the evidence relating to climatic relationships in man, this review has purposely avoided the complexities of cultural adjustment to climate, and it has concentrated on physical characters that are measurable, particularly on morphological variation. This is not to imply that cultural adjustments are negligible; with the development of culture, man places increasingly efficient barriers between himself and his environment. The most primitive of tool users was able to remove his diet from the whim of nature; the first to don an animal skin removed his microclimate from the outside environment. Today we control to some degree our own body shape by choosing our diet, and we eliminate climatic stress by central heating or air conditioning. All cultures intervene to some extent between man and his environment, so that for him "environmental" must assume a connotation somewhat different from that applied to lower creatures. But those features of his culture which are capable of interfering with his climatic environment are, by comparison with the duration of human evolution, very limited in extent, and hence they can have been operative as modifiers of climatic selection for only a relatively brief period. Moreover, for the majority of mankind today there are features of climate that are quite impossible to control. It is for these reasons that the associations here demonstrated are detectable despite man's culture.

It was first argued, from other fields of biological study, that relationships are to be expected between human physical characters and environmental features, and a procedure for their elucidation was set out. In the first part, "The Association with Climate of Adult Physique," a possible mechanism relating variations in body form to environmental demands was explored in a study of basal metabolism, in which a climatic association was found and a hypothesis relating body size and shape to climatic variation was proposed. This hypothesis was validated by the results of examination of variations in adult size and shape, including body weight, length of lower and upper limbs and their segments, and trunk girth and breadths. It was shown that these variations are, with minor modifications, compatible with the rules of Bergmann and Allen, the functional basis of which concerns facilitation of the maintenance of body heat balance.

The physiology of adjustment to heat and cold was examined next. Morphological variations were shown to be of potential significance in survival under climatic stress, and the physiological evidence was found to be compatible with the above hypothesis. From more limited evidence, ethnic differences in the ability to withstand stress of heat and cold emerged, and differences in morphology and body composition in turn contributed to these differences. Indirect effects of physique variation, for instance, through improved performance in difficult conditions, may also be involved in the climatic relationships.

The second part, "Developmental Aspects of the Association," explored how the association of adult physique with climate comes about. The climatic relationship of body shape is discernible in childhood. It indeed appears to be foreshadowed at the moment of birth, in the weight of newborns, though here the mechanism of the climatic association, not yet understood, is felt to be different. As regards plasticity of body form, it appears that nutritional variation exerts only a minor effect on the extremes of tropical physique. Finally, the possibility of a direct effect of climate on the processes of growth, either the period over which growth occurs or the rate of growth during that period, was examined. The evidence suggested relatively little direct climatic influence on growth. It was concluded that while physiological phenotypic modification and genetic differences in combination probably account for the association of adult physique with climate, the evidence suggests that genetic differences account for the major part of the association, and that natural selection acting on population gene pools is ultimately responsible.

The argument that the differences arise from long-term selection would be considerably strengthened were other climatic associations detectable, particularly if they involved mechanisms other than the maintenance of the body heat balance. The third part, "Other Climatic Associations," concerned a miscellany of characters. The evidence for an association of nasal shape with climate is strong, and here the proper humidification of inspired air appears to be the predominating mechanism. For face and head form the immediate selective advantage appears more local but of wide implication, the avoidance of damage to local surface tissues of the face and the maintenance of optimal temperature for brain function. For variation in skin color several distinct mechanisms appear to be involved. The occurrence of pigmentation appears to be related to protection against the deleterious effects of ultraviolet radiation, particularly its carcinogenic effects, though pigmentation may also be related to heat exchange, either directly or by preventing damage to the sweat glands and hence impairment of the heat-regulating mechanism. Lack of pigmentation appears to be related to improvement of synthesis of vitamen D in conditions where its dietary supply and the availability of radiation for its conversion are limited.

These various phenomena lend each other mutual support. If there is climatic selection for one complex of characters to meet one type of climatic demand, then the occurrence of other complexes to meet different climatic requirements is the more likely. It is unfortunate that knowledge of how the associations come about, and in particular of the genetic basis of the characters, is so incomplete, for with this knowledge the rate of evolutionary change could be better measured. Moreover, even though variation of a character within a population may have a genetic basis, this does not mean that variation between populations in this character is also genetic. Much more work remains to be done in order to validate the selective basis of the hypothesis here outlined.

The characters considered here are those in which there has been some attempt at functional validation of the postulated advantages by experimental procedures. There are a number of other characters for which there has been speculation about possible climatic significance but for which there is little experimental evidence, even though clines have been suggested for some of these characters, for example, by Newman for the shape of the face in New World peoples. The shape of the lips of different peoples, the color of the eyes and the hair, the distribution, form, and amount of body and face hair, prognathism, all are distinguishing characters of different populations. But prognathism and lip shape have barely been measured. There is still very little quantitative information on the amount of hair in different populations or on the variations in its body distribution. If the peppercorn hair of Africans provides useful insulation against solar rays without impeding the evaporation of sweat, and the blanket of heavy straight hair insulates the head surface in the Arctic, clines should perhaps be detectable in these as in the characteristics of physique, but again there is very little quantitative information on variations in hair form in which these can be sought. This perhaps is the next task, the quantitation of these hitherto intractable characters and examination of variations in the resulting measures.

Finally, it has been plausibly argued by Weiner (1971) that man, the present lord of creation, may owe his position of superiority to his ancestral climatic adaptation. The cradle of human evolution was undoubtedly the tropics. This is shown both by the evidence of human paleontology and by man's present anatomical and physiological characters. By comparison with his nearest relatives, the anthropoids, the amount of his body hair is reduced. This reduction allows the rapid loss of heat when the body heat load rises to high levels, a loss that is brought about by sweating, and sweating is much less effective if there is a blanket of hair immobilizing an air layer next to the body. Second, man has highly efficient eccrine sweat glands distributed over the whole body, which are capable of producing sweat rapidly. These are almost unique among the higher animals, in which, including man's nearest relatives the apes and monkeys, a histologically distinct type, the apocrine gland, is more usual. In the early tropical days of human evolution there occurred relative loss of hair and replacement of the apocrine glands by the eccrine series, two features retained in common by all human populations today irrespective of where they live. These features give a well-acclimatized individual a very high cooling capacity.

But this efficiency of heat loss has other implications. The human form has a high proportion of muscle mass and lung volume relative to its weight; the capacity of the lungs and heart allows oxygen intake to be greatly increased over the resting requirements and to be efficiently distributed to muscles. The structure and fuel supply are there for a high work output. The efficiency of the cooling system means that work can be long sustained. Hence the human body can not only perform intensive work, it can also continue to work over long periods. Thus was early man able to carry out the demanding tasks of hunting—long-sustained muscular effort in tracking and stalking prey, intensive bursts of high activity in the attack or the final stage of running down game.

It was this cooling efficiency, evolved for its selective advantage in hot conditions, that allowed man's ancestors also to exploit the non-tropics. The physical demands made by primitive hunting can be translated to later cultural phases. The energy expenditure of Neolithic farmers breaking new ground, of Iron Age colonizers clearing new forest, of medieval warriors, of today's lumberjacks, follows a similar pattern. These activities in their turn must have produced selection for larger, heavier, stronger physiques and have contributed to the development and maintenance of human social life. Without these and other activities, allowed by the climatic pre-adaptation, the cultural advances of human history which brought mankind to his present dominating position could not have been achieved.

SUMMARY TABLE: COEFFICIENTS OF CORRELATION WITH TEMPERATURE

	A. Overall relationships		B. Within-variety relationships	
Mean character	Total	Partial	Total	Partial
<u>Basal metabolic rate</u>				
Cal/24 hr	-.736[c]		-.828[c]	
Cal/24 hr[d]	-.845[c]			
Cal/kg	-.430[a]		-.329	
Cal/kg[d]	-.644[b]			
Cal/m^2	-.644[c]		-.528[b]	
Cal/24 hr (2nd series)	-.773[b]	-.679[b,e]		
Weight	-.573[a]			
Stature	-.065			
<u>Weight</u>				
Weight	-.600[c]	-.538[c,f]	-.362[c]	-.352[c,f]
Stature	-.351[c]	+.164[g]	-.166	-.160[g]
Stature^3	-.360[c]		-.163	
Weight[d]	-.809[c]	-.704[c,f]		
Stature[d]	-.587[c]	+.208[g]		

[a] Significant at 5%.
[b] Significant at 1%.
[c] Significant at 0.1%.
[d] Female.
[e] Excluding stature and weight.
[f] Excluding stature.
[g] Excluding weight.

	A. Overall relationships		B. Within-variety relationships	
Mean character	Total	Partial	Total	Partial
Lower limb				
Sitting height	-.428[c]			
Rel. sitting height	-.619[c]		-.386[c]	
Rel. sitting height[d]	-.654[c]		-.403[b]	
Rel. symphysial height	+.426			
Rel. navel height	+.676[b]			
Rel. leg length	+.881[c]			
Rel. trochanteric height	+.590[a]			
Rel. calf circumference	-.558[a]			
Rel. foot length	+.532[b]			
Rel. calf length	+.338[a]			
Calf/thigh index	+.479[a]			

(continued on p. 96)

	A. Overall relationships		B. Within-variety relationships	
Mean character	Total	Partial	Total	Partial
Upper limb				
Span	+.235[b]	+.484[c,f]	+.033	+.260[b]
Stature	+.024		-.078	
Rel. span	+.470[c]			
Rel. biacromial breadth	-.577[c]			
Rel. arm length	+.680[c]			
Rel. upper arm length	+.370[a]			
Rel. forearm length	+.705[c]			
Rel. hand length	+.434[a]			
Antibrachial index	+.586[c]			
Trunk				
Chest girth	-575[c]	-.585[c,f]		
Stature	-.010			
Rel. bicristal breadth	-.823[c]			
Other physiological characters				
Systolic pressure	-.409[b]			
Systolic pressure basal	-.874[c]			

	A. Overall relationships		B. Within-variety relationships	
Mean character	Total	Partial	Total	Partial
Diastolic pressure basal	-.591[a]			
Respiration basal	+.486[a]			
Pulse nonbasal	+.517[b]			
Developmental features				
Log weight in children:				
Age 6 male	-.740[c]	-.742[c,f]		
female	-.659[c]	-.676[c,f]		
Age 10 male	-.631[c]	-.690[c,f]		
female	-.522[c]	-.551[c,f]		
Age 14 male	-.577[c]	-.581[c,f]		
female	-.444[c]	-.401[c,f]		
Birth weight	-.763[c]		-.378[c]	
Menarcheal age	-.567[c]		-.40[a]	

In this table it is the general trends of the correlation coefficients that are important, not their actual values; the coefficients cannot be directly compared with each other, since the samples of data on which they are based are not the same for each character.

For Further Reading

Adam, J. M., F. P. Ellis, R. T. John, T. S. Lee, and R. K. MacPherson (1953). "Periodic and diurnal variations in physiological values for acclimatized naval ratings working at high temperatures for 5 months in the tropics." *Medical Research Council*, Royal Naval Personnel, Report 53/750.

Adolph, E. F. (1947). *Physiology of Man in the Desert*. New York: Interscience.

Allen, B. C. (1961). "Applied anatomy of paranasal sinuses." *Journal of the American Osteopathological Association*, 60: 978-983.

Allen, J. A. (1877). "The influence of physical conditions on the genesis of species." *Radical Review*, 1: 108-140.

Bainbridge, D., and D. F. Roberts (1966). "Dysplasia in Nilotic physique." *Human Biology*, 38: 251-278.

Baker, P. T. (1955). "Relationship of desert heat stress to gross morphology." *Environmental Protection Division Technical Report*, EP7 Q.M.R. & D. Centre.

Baker, P. T. (1958). "The biological adaptation of man to hot deserts." *American Naturalist*, 92: 337.

Baker, P. T. (1966). "Ecological and physiological adaptation in indigenous South Americans." In P. T. Baker and J. S. Weiner, eds., *The Biology of Human Adaptability*, pp. 275-303. Oxford.

Baker, P. T. (1969). "Human adaptations to high altitude." *Science*, 163: 1149-1156.

Barnicot, N. A. (1959). "Climatic-factors in the evolution of human populations." Cold Spring Harbor Symposium on Quantitative Biology, 24: 115

Bar Or, O., H. M. Lundgren, and E. R. Buskirk (1969). "Heat tolerance of exercising obese and lean women." *Journal of Applied Physiology*, 26: 403.

Bass, D. E., C. R. Kleeman, M. Quinn, A. Henschel, and A. H. Hegnauer (1955). "Mechanisms of acclimatization to heat in man." *Medicine*, 34: 323-380.

Benzinger, T. H. (1969). "Heat regulation; homeostasis of central temperature in man." *Physiological Reviews*, 49: 671.

Bergmann, C. (1847). "Uber die Verhältniesse der Wärmeokonomie der Thiere zu ihrer Grösse." *Göttingen Studien*, 1: 595-708.

Blum, H. F. (1945). "The physiological effects of sunlight on man." *Physiological Review*, 25: 483.

Blumenbach, J. F. (1865). Anthropological Treatises, trans. T. Bendyshe, London.

Boas, F. (1910). *Changes in Bodily Form of Descendants of Immigrants*. Government Printing Office, Washington, D.C.

Buskirk, E. R., H. Lundgren, and L. Magnusson (1955). "Heat acclimatisation patterns in obese and lean individuals." *Annals of the New York Academy of Sciences*, 131: 637.

Carlson, L. D., and A. C. L. Hsieh (1965). In O. G. Edholm, ed., *The Physiology of Human Survival*, 15. New York: Academic Press.

Carter, J. E. L. (1970). "The somatotypes of athletes." *Human Biology*, 42: 535-569.

Collins, K. J., and J. S. Weiner (1965). "The effect of heat acclimatization on the activity and number of sweat glands." *Journal of Physiology*, 177: 161-171.

Collins, K. J., and J. S. Weiner (1968). "Endocrinological aspects of exposure to high environmental temperatures." *Physiological Review*, 48: 785-839.

Coon, C. S., S. M. Garn, and J. B. Birdsell (1950). *Races*. Springfield, Ill.: Charles C. Thomas.

Darwin, C. (1871). *The Descent of Man*. London.

Darwin, C. (1875). *The Origin of Species*. London.

Davies, A. (1932). "A resurvey of the morphology of the nose in relation to climate." *Journal of the Royal Anthropological Institute*, 62: 337-359.

deMüller, J. W. (1853). *Causes de la Coloration de la Peau*. Stuttgart.

Dobzhansky, T. (1962). *Mankind Evolving*. New Haven, Conn.: Yale University Press.

Eckert-Mobius, A. (1933). "Vergleichend anatomisch-physiologische studie uber sinn und Zweck der Nasennebenhohlen des Menschen und der Saugetiere." *Archiv fur Ohren-Nasen und Kehlkopfheilkunde*, 134: 288-307.

Edwards, E. A., and S. Q. Duntley (1939). "The pigments and colour of living human skin.: *American Journal of Anatomy*, 65: 1.

Edwards, M., and A. C. Burton (1960). "Temperature distribution over the human head, especially in the cold." *Journal of Applied Physiology*, 15: 209.

Egan, C. J. (1963). "Local vascular adaptations to cold in man." *Federation Proceedings*, 22: 547-551.

Elsner, R. W., K. L. Andersen, and L. Hermansen (1960). "Thermal and metabolic responses of Arctic Indians to moderate cold exposure at the end of winter." *Journal of Applied Physiology*, 15: 659-661.

Ewert, G. (1965). "On the mucus flow rate in the human nose." *Acta Otolaryngologica*, Stockholm, Supplement, 200: 1-62.

Froese, G., and A. C. Burton (1957). "Heat losses from the human head." *Journal of Applied Physiology*, 10: 235.

Garn, S. N. (1961). *Human Races*. Springfield, Ill.: Charles C. Thomas.

Gloger, C. L. (1833). *Das Abändern der Vogel durch Einfluss des Klimas*. Breslau.

Goldstein, M. (1953). "Demographic and body changes in descendants of Mexican immigrants." Institute of Latin American Studies, University of Texas, Austin.

Greulich, W. W. (1957). "A comparison of the physical growth and development of American born and native Japanese children." *American Journal of Physical Anthropology*, 15: 489-516.

Hammel, T. H. (1964). "Terrestrial animals in cold; recent studies of primitive man." In D. B. Dirrel et al., *Handbook of Physiology*, "Adaptation of the Environment." *American Physiological Society*, Washington, D.C.

Hanna, J. M. (1970a). "Responses of native and migrant desert residents to arid heat." *American Journal of Physical Anthropology*, 32: 187-196.

Hanna, J. M. (1970b). "A comparison of laboratory and field studies of cold response." *American Journal of Physical Anthropology*, 32: 227-232.

Hanna, J. M., and P. T. Baker (1974). "Comparative heat tolerance of Shipibo Indians and Peruvian Mestizos." *Human Biology*, 46: 69-80.

Hardy, J. D. (1961). "Physiology of temperature regulation." *Physiological Reviews*, 41: 521.

Harrison, G. A., J. S. Weiner, J. M. Tanner, and N. A. Barnicot (1964). *Human Biology*. Oxford.

Hulse, F. S. (1957). "Some factors influencing the relative proportions of human racial stocks." *Cold Spring Harbour Symposium, Quantitative Biology*, 22: 33-45.

Hulse, F. S. (1968). "Migration and cultural selection in human genetics." *The Anthropologist*, special volume, Delhi.

Hulse, F. S. (1971). *The Human Species*. New York: Random House.

Huxley, J. (1942). *Evolution, the Modern Synthesis*. London.

Iampietro, P. F., R. F. Goldman, E. R. Buskirk, and D. E. Bass (1959). "Responses of Negro and white males to col." *Journal of Applied Physiology*, 14: 798-800.

Irving, L., K. L. Andersen, A. Bolstad, et al. (1960). "Metabolism and temperature of Arctic Indians during a cold night." *Journal of Applied Physiology*, 15: 635-637.

Itoh, S., H. Shirato, and A. Kuroshima (1968). "Cold pressor response of the Ainu." *Journal of the Physiological Society of Japan*, 30: 49-50.

Ivanowsky, A. (1925). "Die anthropometrischen Veränderungen russischer Völker unter den Einfluss der Hungersnot." *Archiv fur Anthropologie*, 20: 1.

Kerslake, D. M. (1972). *The Stress of Hot Environments*. Cambridge University Press.

Knip, A. S. (1975). "Acclimatization and maximum number of functioning sweat glands in Hindu and Dutch females and males." *Annals of Human Biology*, 2: 261-277.

Koertvelyessy, T. (1972). "Relationships between frontal sinus and climatic conditions." *American Journal of Physical Anthropology*, 37: 161-172.

Kollias, J., L. Bartlett, V. Bergsteinova, J. S. Skinner, E. R. Buskirk, and W. C. Nicholas (1974). "Metabolic and thermal responses of women during cooling in water." *Journal of Applied Physiology*, 36:577-580.

Kondo, S. (1969). "A study on acclimatization of the Ainu and the Japanese with reference to hunting temperature reaction." *Journal of the Faculty of Science*, Tokyo University, section 5, 3: 253-265.

Krog, J., and M. Wika (1971-72). "Studies of the peripheral circulation in the hand of the Igloolik Eskimo." In "International Biological Program, Human Adaptability Project (Igloolik, N.W.T.)" Annual Report No. 4: 173-181. Anthropological Series: No. 11, University of Toronto.

Ladell, W. S. S. (1951). "Assessment of group acclimatization to heat and humidity." *Journal of Physiology*, 115: 296-312.

Ladell, W. S. S. (1964). "Terrestrial animals in humid heat: man." In D. B. Dill, E. F. Adolph, and C. G. Wilber, eds., *Handbook of Physiology*, pp. 625-659.

Lasker, G. W. (1946). "Migration and physical differentiation: a comparison of immigrants with American born Chinese." *American Journal of Physical Anthroplogy*, 4: 273-300.

Lasker, G. W. (1952). "Environmental growth factors and selective migration." *Human Biology*, 24: 262-289.

Lasker, G. W. (1954). "The question of physical selection of Mexican immigrants to the U.S.A." *Human Biology*, 26: 52-57.

Lind, A. R., P. W. Humphreys, K. J. Collins, K. Foster, and K. F. Swetland (1970). "Influence of age and daily duration of exposure on responses of men to work in heat." *Journal of Applied Physiology*, 28: 50-56.

Little, M. A., R. B. Thomas, and J. W. Larrick (1973). "Skin temperature and cold pressor responses of Andean Indians during hand immersion in water at 4°C." *Human Biology*, 45: 643-662.

Lund-Larsen, K., M. Wika, and J. Krog (1970). "Circulatory Responses of the hand of Greenlanders to local cold stimulation." *Arctic Anthropology*, 7(1): 21-25.

MacPherson, R. K., ed. (1960). "Physiological responses to hot environments." Medical Research Council special report series no. 298.

Marrett, J. F. (1936). *Race, Sex and Environment*. London: Hutchinsons.

Maurer, R. (1953). "Zur Physiologie der Schadelpneumatisation." *Archiv fur Ohren-Nasan und Kehlkopfheilkunde*, 163: 471-473.

Mayr, E. (1963). *Animal Species and Evolution*. Cambridge, Mass.: Harvard University Press.

Meehan, J. P. (1955). "Individual and racial variation in a vascular response to a cold stimulus." *Military Medicine*, 116: 330-334.

Meredith, H. V. (1970). "Body weight at birth of viable human infants: a worldwide comparative treatise." *Human Biology*, 42: 217-264.

Negus, V. (1957). "The function of paranasal sinuses." *Archives of Otolaryngology*, 66: 430-442.

Newman, M. T. (1953). "The application of ecological rules to the racial anthropology of the aboriginal new world." *American Anthropologist*, 55: 311.

Newman, M. T. (1956). "Adaptation of man to cold climates." *Evolution*, 10: 101-105.

Newman, M. T. (1960). "Adaptations in the physique of American aborigines to nutritional factors." *Human Biology*, 32: 288.

Newman, R. W. (1969). "Cold acclimation in Negro Americans." *Journal of Applied Phsyiology*, 27: 316.

Newman, R. W., and E. H. Munro (1955). "The relation of climate and body size in U.S. males." *American Journal of Physical Anthropology*, 13: 1.

Orr, K. D., and D. C. Fainer (1952). "Cold injuries in Korea during the winter of 1950-51." *Medicine*, 31: 177.

Paik, K. S., B. S. Kang, D. S. Han, D. W. Rennie, and S. K. Hong (1972). "Vascular responses of Korean ama to hand immersion in cold water." *Journal of Applied Physiology*, 32: 446-450.

Quenouille, M. H., A. W. Boyne, W. B. Fisher, and I. Leitch (1951). "Statistical studies of recorded energy expenditure of man." Commonwealth Bureau of Animal Nutrition, technical communication no. 17.

Raven, P. B., S. Taguchi, B. L. Drinkwater, M. Kaneko, S. M. Horvath, and H. Matsui (1974). "Anthropometric, spirometric, and physiologic comparisons of migrant Japanese." *Human Biology*, 46: 483-494.

Ridgeway, W. (1908). "The application of zoological laws to man." Report of the British Association for the Advancement of Science, pp. 832-847.

Roberts, D. F. (1952a). "An ecological approach to physical anthropology." Proceedings of the 4th International Congress of Anthropological Sciences, Vienna, p. 145.

Roberts, D. F. (1952b). "Basal metabolism, race, and climate." *Journal of the Royal Anthropological Institute*, 82, part ii: 169-183.

Roberts, D. F. (1953). "Body weight, race, and climate." *American Journal of Physical Anthropology*, 11 (n.s.), 4: 533-558.

Roberts, D. F. (1960). "Growth studies of African children." In J. M. Tanner, ed., *Human Growth*, pp. 59-72. New York: Pergamon Press.

Roberts, D. F. (1961). "Korperhöhe und Gewicht nilotider Kinder." *Homo*, 12: 33-41.

Roberts, D. F. (1969). "Race, genetics, and growth." *Journal of Biosocial Science Supplement* 1: 43-67.

Roberts, D. F. (1975). "Environment and the fetus." In D. F. Roberts and A. M. Thomson, eds., *The Biology of Human Fetal Growth*. London: Taylor and Francis.

Roberts, D. F., and D. Bainbridge (1963). "Nilotic physique." *American Journal of Physical Anthropology*, 21: 341-370.

Roberts, D. F., and D. P. S. Kahlon (1976). "Environmental correlations of skin colour." *Annals of Human Biology*, 3: 11-22.

Roberts, D. F., F. M. Salzano, and J. O. C. Willson (1970). "Active sweat gland distribution in Caingang Indians." *American Journal of Physical Anthropology*, 32(3): 395-400.

Robinson, S. (1942). "The effect of body size upon energy exchanges in work." *American Journal of Physiology*, 136: 363.

Schickele, E. (1947). "Environment and fatal heat stroke; an analysis of cases occurring in the Army in the U.S. during World War II." *Military Surgeon*, 100: 235-256.

Scholander, P. F., T. H. Hammel, J. S. Hart, D. H. LeMessurier, and J. Steen (1958). "Cold adaptation in Australian aborigines." *Journal of Applied Physiology*, 13: 211-218.

Scholander, P. F., K. Lange Andersen, J. Krog, F. Vogt Lorentzen, and J. Steen (1957). "Critical temperature in Lapps." *Journal of Applied Physiology*, 10: 231.

Schreider, E. (1953). "Variability of internal environment and the mechanisms of biological correlations." *Nature*, 171: 339.

Schumann, L. M. (1951-1952). "Epidemiology of frost bite." Army Medical Research Laboratory Report, p. 113.

Shapiro, H. (1939). *Migration and Environment*. London.

Simpson, G. (1950). *The Meaning of Evolution*. London.

Smith, S. Stanhope (1810). Essay on the Causes of the Variety of Complexion and Figure in the Human Species. New Brunswick.

Steegmann, A. T. (1972). "Cold response, body form, and craniofacial shape in two racial groups of Hawaii." *American Journal of Physical Anthropology*, 37: 193-222.

Steegmann, A. T. (1974). "Ethnic and anthropometric factors in finger cooling." *Human Biology*, 46: 621-631.

Suski, P. M. (1933). "The body build of American born Japanese." *Biometrics*, 25: 323-352.

Suzuki, M. (1969). "Peripheral response to cold." *Journal of the Anthropological Society of Nippon*, 77: 213-223.

Tanner, J. M. (1962). *Growth at Adolescence*. Oxford.

Thies, R. E., J. R. Billinghurst, and H. D. Richardson (1967). "Motor nerve conduction velocities in healthy young east Africans." *Journal of Applied Physiology*, 23: 321.

Thomson. A. (1913). "The correlations of isotherms with variations in the nasal index." International Congress of Medicine, London, 17: 89-90.

Thomson, A., and L. D. Buxton (1923). "Man's nasal index in relation to certain climatic conditions." *Journal of the Royal Anthropological Institute*, 53: 92.

Thomson, M. L. (1954). "A comparison between the number and distribution of functioning eccrine sweat glands in Europeans and Africans." *Journal of Physiology*, 123: 225-233.

Wagner, J. A., S. Robinson, S. P. Tzanoff, and R. P. Manio (1972). "Heat tolerance and acclimatization to work in the heat in relation to age." *Journal of Applied Physiology*, 33: 616-622.

Webb, P. (1951). "Air temperatures in respiratory tracts of resting subjects in cold." *Journal of Applied Physiology*, 4: 378.

Weiner, J. S. (1950). "Observations on the working ability of Bantu mine workers." *British Journal of Industrial Medicine*, 7: 17-26.

Weiner, J. S. (1954). "Nose shape and climate." *American Journal of Physical Anthropology*, 12: 1-4.

Weiner, J. S. (1967). "Aspects of adaptation." *Advancement of Science*, 23: 1-9.

Weiner, J. S. (1971). *Man's Natural History*. London: Weidenfeld and Nicolson.

Williams, D. D., J. H. Petajan, and S. B. Lee (1969). "Summer and winter hand calorimetry of children in the village of Old Crow." *International Journal of Biometeorology*, 13 (part II): 11.

Wyndham, C. H. (1970). "Adaptation to heat and cold." In D. H. K. Lee and D. Minard, eds. *Physiology, Environment, and Man*. New York: Academic Press.

Wyndham, C. H., W. Bonner, M. G. Devine, and W. E. Paterson (1952). "Physiological responses of African labourers at various saturated air temperatures, wind velocities, and rates of energy expenditure." *Journal of Applied Physiology*, 6: 290-298.

Wyndham, C. H., and H. Loots (1969). "Responses to cold during a year in Antarctica." *Journal of Applied Physiology*, 27: 696.

Yoshimura, H., and T. Iida (1952). "Studies on the reactivity of skin vessels to extreme cold." *Japan Journal of Physiology*, 2: 177-185.

Author Index

Subject Index